YOUR NUMBER'S UP

a calculated approach
to successful math study

YOUR NUMBER'S UP

a calculated approach
to successful math study

C. Ann Oxrieder
Janet P. Ray
Seattle Community College

ADDISON-WESLEY PUBLISHING COMPANY
Reading, Massachusetts • Menlo Park, California
London • Amsterdam • Don Mills, Ontario • Sydney

Library of Congress Cataloging in Publication Data

Oxrieder, C. Ann, 1946–
    Your number's up.

    Bibliography: p.
    Includes index.
    1. Mathematics–Study and teaching–Psychological
aspects.   I. Ray, Janet P., 1943–   II. Title.
QA11.096   510'.7'1   81-19153
ISBN 0-201-05526-0   AACR2

ISBN 0-201-05526-0
ABCDEFGHIJ-AL-898765432

PREFACE

The motivation for our interest in math anxiety came from two directions. On the one hand it began with our concern that as a whole, women were locking themselves out of the vast majority of career options by choosing to take very little math.

This society is now in a period where there are more and more job opportunities in technical or related fields. These are traditionally areas for which women have the least preparation. In our minds, it is critical that women receive the support and undertake the education essential to take advantage of these opportunities.

On the other hand, we had worked with students, both men and women, who were experiencing extreme distress while studying math. We found it difficult to understand why otherwise competent students often experienced difficulty in math classes. Was there something peculiar about math that made learning it different and more difficult than learning other subjects? Or was the problem that students were entering math classes with attitudes and expectations that made success difficult?

We concluded that no matter what the causes, any condition that prevented a person from reaching his or her goals was worthy of attention.

Both groups we were interested in could benefit from a class that presented math learning skills in a supportive environment. So we developed a course designed to help reduce math anxiety. Many women enrolled and so did some men. We found out that although much of the literature spoke of math anxiety or avoidance as a "women's problem," men were affected in significant numbers as well.

The course was based on the philosophy that students would benefit from the development of self-management skills for coping with anxiety in addition to doing work on math-related skills. This book is a reflection of that philosophy. It contains both counseling information and math activities, and it emphasizes practical skill development—study skills, communication skills, stress-control skills, as well as problem-solving skills.

We do not attempt to provide instruction in any particular aspect of mathematics. That omission is intentional. We found math-anxious individuals at all levels of mathematics from the most elementary to the more advanced. We attempted to create a book that would be helpful no matter what particular mathematical subject is being studied.

This book is intended for use by anyone wishing to pursue their involvement with mathematics further. It is appropriate for math students from high school through college levels as a supplement to course textbooks, for adults who anticipate having to take a math course sometime in the future, as a textbook for courses dealing with the reduction of math anxiety, and as a resource for mathematics teachers at any level who work with students exhibiting math anxiety.

There is an instructor's guide available for those of you who may be using this book as a resource in a math anxiety reduction program. It contains further details of our instructional strategies as well as additional exercises and classroom activities.

The challenge of overcoming math anxiety is a tremendous one. Yet we feel efforts that individuals make in this direction are worthwhile. The satisfaction that is felt is great when even small steps are made in confronting longstanding fears. We hope this book may be of some assistance in helping people to complete successfully those small steps that can add up to significant change.

ACKNOWLEDGMENTS

We would like to acknowledge our indebtedness to the many people who share the responsibility for the origin and completion of this book. They include

- Jean Smith, Bonnie Donady, and Jerine Ridgway, presentors of a workshop on math anxiety at the University of Washington, that inspired us to develop a course called "Math without Fear" and ultimately this workbook;

- Students at Seattle Central Community College who confronted their math anxiety by enrolling in our course and who helped field test much of the material included here;

- Betty Richardson who read and reread our material from the point of view of the math-anxious reader, and who suggested countless improvements in the text;

- Sylvia Soholt and Liana Nakamura for graphics on the final manuscript;

- Mary Mahler, our patient typist, who read the unreadable and put it together properly;

- Suzanne Shafer, who proofread (repeatedly) our first manuscript and who drew the original mathematical illustrations;

- The many dedicated and concerned mathematics instructors we spoke with who recognized the existence of math anxiety and who encouraged our efforts;

- Seattle Community College District, which funded our first attempts at developing curriculum materials;

- Our reviewers—James W. Daniel, University of Texas at Austin; Linda D. Falstein, University of Massachusetts/Boston; Helen R. Santiz, University of Michigan–Dearborn; Jean Smith, Middlesex Community College; Abby Tanenbaum; and Mary Walz-Chojnecki, University of Wisconsin–Milwaukee—whose thoughtful suggestions have enhanced and expanded our work. Special thanks to our final reviewers, for their careful attention to detail and accuracy; and

- Patricia Mallion, of Addison-Wesley, who encouraged us to take our efforts seriously and who provided the support necessary to make this a reality.

CONTENTS

**what is it and how
do I know if I have it?**

1

MATH
ANXIETY
DEFINED

This chapter will cover

- *a definition of math anxiety,*
- *various signs of math anxiety,*
- *reasons for math anxiety, and*
- *your math history.*

DEFINITION OF MATH ANXIETY

The dictionary defines anxiety as a "state of being uneasy . . . or worried about what may happen." Many people who avoid math will say that being uneasy or worried is much too mild to describe their feelings when called upon to do math. When faced with mathematical tasks, many people experience discomfort ranging from mild tension to extreme mental and physical suffering. Robinson and Sims in a study done in 1972 have defined math anxiety as "feelings of tension and anxiety that interfere with the manipulation of numbers and the solving of problems in a wide variety of ordinary life and academic situations."

SIGNS OF MATH ANXIETY

Common reactions of math anxiety include the following:

- **Blocking Out** Suddenly everything goes blank. You forget the math skills you know. You may be unable to understand the problem, to do even the simplest addition, or to decide what to do next. It is as if a large curtain is drawn down separating you from all your knowledge and abilities.

- **Tension** You may begin working the problem but feel your body tightening up. Your neck or back is stiff, your hand may shake, your breathing may become strained.

- **Panic** There is a feeling of coming disaster. Your pulse races and you perspire heavily. You're sure the problem cannot be done; your feelings of defeat take control.

- **Paranoia** You suspect that everyone knows how stupid you're feeling. You think, "It's an easy problem; probably everyone in the room can do it except me."

- **Tune Out** When numbers come up or math is mentioned, the rest of the conversation is lost on you. You fail to hear what the person is saying. If you're reading a book, at the end of the page you have no idea what you have read.

- **Guilt** You may feel that even your ability in the little bit of math you can do is a fraud. You've been faking what you can do and sooner or later the deception will be discovered. Maybe you suspect that it's really your fault you have the problem.

- **Physical Reactions** When you deal with math you may get a headache, become nauseous, have stomach cramps, experience blurred vision, lose your ability to concentrate, or get very sleepy.

- **Avoidance** Sometimes the most comfortable way to deal with a problem is to avoid it. You may try hard to stay out of situations where you encounter math. Avoiding math may also be a sign of math anxiety.

Do any of these reactions sound familiar to you? Most people, even mathematicians, experience some of them some of the time. These reactions are usually stronger in stressful situations, like taking tests and being called on in class. When we'd like to do our best, when the pressure is on, we experience more of the symptoms of math anxiety. One of the results of these reactions is that we perform poorly and this adds to the embarrassment. So the next time we have to deal with mathematics, the problem is even larger and the cycle builds.

What kinds of people are affected by these reactions? All kinds. Much of the early work on math anxiety was done by educators and others seeking to explain why women did not choose majors in the scientific and technical fields. As a result, math anxiety is often talked about as a woman's problem. Further research has shown that men are probably just as frequently affected although they may deal differently with their feelings about it. In addition, educational level does not seem to be a significant factor. The holder of a Ph.D. in history is as likely to fear math as is a high school dropout. As was mentioned earlier, even mathematicians report experiencing symptoms on occasion.

REASONS FOR MATH ANXIETY

The origins of any strong feeling are very complex. There is no single reason why many people fear math. So when this book lists causes, it is only suggesting some of the factors that may lead to math anxiety.

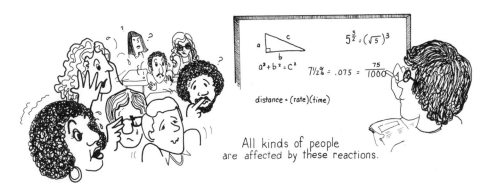

All kinds of people
are affected by these reactions.

It is listing the key reasons given by people who have been suffering from math anxiety or math avoidance. It may be helpful both to those who dislike mathematics and to those who stay away from math to hear what these people have to say. It is not important whether the events cited are actual causes or not. How people view a situation is the critical factor in how they behave. What follows represents many different experiences and points of view.

*I remember clearly. It was in second grade when we were
doing fractions. I just got stuck and I was scared to death
from then on of being found out.*

The age at which math anxiety begins varies widely. Some can recall intense negative reactions as early as the first grade in school. For others math was a favorite subject until a much later time in their lives. Often feelings changed when they encountered a particular branch of mathematics like algebra or geometry. Significantly, very few people say they had negative experiences in mathematics before starting school. It is likely that for many victims the cause of the difficulty is school. This is unfortunate since school is the very institution that is supposed to be encouraging in us a love of learning—including a love of learning mathematics.

*Mrs. Johnson used to hit me with a ruler if I couldn't
answer the question in front of the class.*

Sometimes the problem can be traced to an unhappy encounter with a particular teacher. This may be someone who either teaches the subject poorly or who treats students unfairly. Instructors who do little to help

students learn math and then scold them for their (predictable) failure, appear in the early histories of far too many math-anxious people. Some teachers make fun of children, or adults, in front of the class for their blunders or lack of success. Others praise only the brightest and quickest students. These both seem like obvious teaching mistakes, yet they occur repeatedly in the classroom. Being singled out and humiliated in a math class has a profound effect on all but the most self-confident. What we often fail to recognize is that many teachers may themselves be math anxious or math haters. It is no wonder that their reactions do not enhance learning. They may be so busy dealing with their own problems that it leaves them unable to cope with student needs. It may also make them intolerant of students who are having problems. In their defense, it should be noted that many teachers, particularly in the early grades, are teaching math against their own better judgment and desires.

> *I lived in constant fear that I would be asked to go to the blackboard and show how to solve a problem. I knew how to do most of them but put me in front of a blackboard and I'd panic.*

For some people, the methods of teaching are the culprits. Working alone at the blackboard when you're having trouble is often overwhelming. Being required to give answers quickly or to compete with other class members can be very uncomfortable. Of course, a certain degree of discomfort accompanies learning any new task, but when the discomfort takes over, learning stops.

Also, being in a class where questions are discouraged or forbidden makes it difficult to clear up uncertainty. You begin to believe that you're stupid if you need to ask questions. Inflexibility in the way problems are done often shuts off creativity in mathematics and gives us an unhealthy distrust of our own intuition. It may also lead us to stop attempting math in angry defense.

> *Oh, I knew how to do the problems, all right, but after a while I refused to do them because they always had to be done the teacher's way.*

Other kinds of pressure, from parents and teachers, can lead a person to hate or fear math. Often math is taught in a very strict environment with an emphasis on precision and right answers that is foreign to other subjects. Instructors who continually allow only one "correct" method (always their way, of course) tend to discourage students. In some cases an individual's decision to avoid math is as much a rebellion against a teacher who is a dictator as it is a dislike of mathematics. In addition, math is often the subject in which parents and others insist that we must do well. We are told it's absolutely essential and we've got to do it, but no effort is made to convince us why. We're not given help to overcome difficulties with the subject and our lack of success is used to make us feel stupid or lazy. Some children are even physically punished for failure in mathematics.

> *We moved to four different states when I was in grade school. They were never doing the same thing in any new class I started. I was lost from arithmetic on.*

Some people can trace difficulty in mathematics to an event that caused them to get behind, like an accident or long illness that meant missed school days. Sometimes moving to a new school meant entering a new class that was doing totally unfamiliar material. Since math is a very cumulative subject, where one topic builds on the next, missing part or all of a previous topic is like trying to erect a brick wall leaving out half of one layer of bricks. You may get by for a couple of more levels, but when too much weight piles up on the top, the whole thing comes crashing down.

*I was okay until they made us use the "new math." Even
the teacher couldn't answer the questions.*

In theory the "new math" curriculum was supposed to make math
more understandable. In many cases it had the opposite effect. Children
who were already competent in basic math skills were now asked to
explain why their answers were true. The shift from learning mechanics
to explaining theory left many students floundering. It was like playing
in a game where suddenly the rules were changed at halftime. In addition,
many teachers of the "new math" were not adequately prepared and
trained in the method. They also felt confused and defensive. Parents
could no longer help with math homework since the material was new to
them also. The result is that many children caught in this period developed a distrust or dislike for math.

*I was told to take home economics because girls didn't
need math and I was having trouble with it anyway so
why waste my time.*

If you are a woman who feels math anxiety, part of the explanation
may center on definitions of masculinity and femininity. These come
from parents, teachers, counselors, peers, and society as a whole. They
say that men are logical and have mathematical minds and that women are
lacking in both qualities. As recently as the last century it was widely
believed that women did not have the mental capacity to concentrate and
reason; their brains were thought to be too small. In addition, they were
considered too illogical and emotional by nature.

Women, like dinosaurs, have
been accused of having brains too small to cope...

Today, on the surface, we scoff at such silliness, but many still question whether women can or should do well in math. Female mathematicians are often regarded with surprise and dismay. Many people still see mathematics as a distinctly male domain. If such people are counselors, they encourage boys to take more math and discourage girls from doing the same. If they are parents, they may insist that a son stick to a math course even if he's having great difficulty, but will allow a daughter to withdraw at the first sign of trouble. If they are teachers, they may encourage and challenge male students while discouraging female students. When a teacher expects male students to do well and female students to do poorly, students often live up to these expectations.

Peer pressure also helps determine how women perform in mathematics and what math courses they choose to take. Since math is often seen as a masculine subject, women who do well in it are made to feel unusual or even unfeminine. Whereas some people side-step math to avoid appearing "too dumb," women often drop out to avoid appearing "too smart." Women are thus provided with a convenient excuse for avoiding math, one that is not readily available to men. If a woman is having difficulty with math, society says that it's okay! It's just part of her nature and besides she'll probably never need it. (For more discussion of this issue, see Chapter 2.)

My mother died just about the time we started percents.
It was more than I could handle all at once.

Finally, for some, the trigger that causes the initial problem in math may have little to do with the subject itself. It may be an unhappy event in one's personal or family life that just comes at a time when math is difficult. Since some parts of mathematics are quite hard to grasp at first and require a lot of concentration, this situation is not unlikely. The strong feelings of grief, frustration, or confusion associated with the outside event become part of the baggage we then hang on to mathematics.

We have discussed some of the issues that lead to people's negative feelings about mathematics and that then affect their ability to perform well in the subject. Now we will ask you to look into the past to come up with a picture of your own math history. Even though it may seem like an unpleasant activity, there are several reasons for doing this. First, it is a good release to describe your feelings about math, to consider where things first went wrong and why. You will feel better once you've recognized and accepted your true feelings about the subject. It's better than appearing to be comfortable doing math when you aren't, or avoiding doing math to save face. Second, you are going to be working on present and future relations with math. It will be useful to recognize the influence of the past as you try to deal with the present.

Some people understand situations more clearly when they write down notes, sentences, paragraphs, or lists and can read them back. Others learn best from a diagram or a picture. We are now going to ask you to use one of these two methods to compose your own math history.

Exercise: **YOUR HISTORY**

Method 1—Written Form

Directions: Answer these questions as best you can. Writing down what happened in the past will help explain some of your present feelings toward math.

1. Do you remember a particular math class where things suddenly started going wrong?

2. Who was the teacher for this class? Can you describe him/her?

3. What math topic(s) were you studying?

4. Were there any other complications in your life at that time? For example, was your family moving around a lot or were you sick and missing school?

5. How did you feel when things started going wrong?

6. What happened after that? Did you try to get out of taking math or did you continue?

7. How do you rate your math ability now?

8. Try to anticipate taking another math class. How do you feel when you think about this?

Method 2—Pictured Form

Directions: Draw a picture of your math history. An example is shown below.

CONCLUSION

Once you have recalled your math history you may want to trade horror stories with other individuals. It will be an immense relief to discover that others too have suffered, and you are not alone. It is important to remember that placing the blame on a specific person or event in the past, although it is a useful and important exercise, does not change the current state of affairs. It is a significant step to be able to admit that you suffered from math anxiety; it is also useful, maybe essential, to understand how and why you got there. But you must be careful that the cause does not become the excuse. Because you feel one way now is not sufficient reason to remain feeling that way. Are you ready now to ask, "What next?" Is it worth the effort that will be necessary to overcome negative attitudes about mathematics? An educator once said, "Just because someone doesn't know something is no reason to teach it to them." You might be tempted to say, "Just because I've been a math avoider is no reason to stop being one now."

The next chapters will provide some information on the advantages of establishing a more comfortable relationship with math. They will offer both new ideas to consider and specific techniques to improve your performance in math situations and reduce your anxiety.

FURTHER READING

Kogelman, Stanley and Joseph Warren. *Mind Over Math*. New York: McGraw-Hill Book Company, 1978.

Tobias, Sheila. *Overcoming Math Anxiety*. New York: W.W. Norton and Company, Inc., 1978.

the myths and misconceptions surrounding mathematics

2

WHAT MATH IS NOT

This chapter will

- *ask you to identify your beliefs about math and mathematicians, and*
- *describe 11 myths that many people believe about math and mathematicians.*

There exists a whole series of popular views about the nature of mathematics and mathematicians. These beliefs affect the way in which we deal with the subject. Some of these beliefs are quite valid. Others could be called myth or exaggeration. While mathematicians, for the most part, probably don't accept these myths, they have done little to dispel them. Perhaps some of us like to believe these ideas because they provide comfortable reasons for us to avoid math. This chapter will examine several of the more commonly held beliefs about mathematics. It will also show how people who have these ideas are affected in their math performance.

Before reading the myths it might be helpful to examine your own beliefs.

Exercise: **YOUR BELIEFS**

Directions: Check the words that best describe your ideas on math.

_____ 1. A subject that some people can do and others just can't

_____ 2. A boring subject

_____ 3. A subject we need for everyday survival

_____ 4. Something only the really smart should take; for the rest it's a waste of time

_____ 5. A rigid, uncreative subject

_____ 6. A subject men do better in than women

_____ 7. A fun subject that is a challenge to learn

_____ 8. A subject that has no connection to the real world

_____ 9. A subject everybody can learn to do well

Now compare your responses with the following. Checks next to items 3, 7, or 9 indicate a generally positive attitude toward math. Checks next to items 1, 2, 4, 5, 6 or 8 indicate feelings of frustration or indifference toward math. They could work against you in your efforts to master the subject. As you read about the myths that follow, try to think further about your own attitudes. Maybe some of them are old beliefs that you could now choose to modify or discard.

THE MATHEMATICAL MIND

This belief holds that some people have the ability to do math easily, while others do not. If you do not have a "math mind" it means somehow that your head just does not function properly. Carried to the extreme, this says there is an inborn characteristic that determines that math will be possible for some folks and impossible for others. The consequences of this myth are disastrous. Clearly if this were true, then no amount of struggle would allow you to succeed in math unless you happen to be born with a "math mind." When you encounter a difficult topic in mathematics it would be all too easy to quit. After all, if you don't have a math mind, what's the use? Even math successes become unimportant when you believe you don't have the necessary math mind because if math is just not your subject, then it is just a matter of time until you'll have problems. Failure is certain.

It may be helpful to relate this kind of attitude to different subjects like history or learning to play tennis. Picture yourself just coming out of a history exam on which you've done badly and saying to a friend, "Well, I just don't have a historical mind." It's much more likely that you would explain the problem by saying you studied the wrong material or didn't understand the questions. You might even say there was too much material to memorize and memorizing is hard for you. You might decide that in the future you would need to spend more time studying, but you wouldn't just give up in despair.

Or imagine that you're a beginning tennis player. In the course of your last practice session you succeeded in hitting a grand total of only six balls over the net, and each of these also went over the court fence into the woods beyond. It's unlikely that you have said that you will never be able to play tennis. We expect to do poorly when we are learning a new skill and expect to spend lots of time before getting good at it.

Yet mathematics seems to be different; when things get tough or when we perform poorly, we're all too ready to give up. We believe we have failed because of some innate deficiency in ourselves over which we have no control. It's probably true that people are not endowed equally with mathematical ability and aptitude. Just as it's true that not everyone has the physical talent to become Billie Jean King or Bjorn Borg. However, most of us can learn to enjoy a reasonable game of tennis, just as most of us can be successful at a reasonable level of mathematics. We may not be an Isaac Newton and develop calculus from scratch, but we can learn to do algebra, apply calculus, understand statistics, or learn to program a computer in BASIC*. This level of mathematics is all that is necessary or relevant to the lives of most of us.

Before you read the next section, take a moment to check off the words that best describe your ideas on what mathematicians are like. Do you think they

_____ 1. have above average intelligence?

_____ 2. rarely make arithmetic mistakes?

---

*BASIC is the name of a language used in computer programming.

_____ 3. can work problems quickly?

_____ 4. think there is only one right method or are fussy people?

_____ 5. are intellectually superior?

_____ 6. are either eccentric or boring?

_____ 7. do well in all school subjects?

THE GENIUS MYTH

Common folklore says that people who are successful in mathematics are smarter (and maybe "better") than the rest of us. (This is one on which mathematicians certainly will not try to change our minds.) We seem to view mathematical ability as higher or more enlightened than other abilities. Individuals who are brilliant in another field, such as music, will be embarrassed if they must admit they find mathematics hard. It is almost as if their weak math skills reflect on their other abilities. It is hard to imagine similar embarrassment if they were to admit that they did not draw well or hadn't learned to tune an automobile properly.

It is important to put mathematical ability on the same level with other skills. There is no greater virtue in being able to do mathematics than there is in being able to write a moving poem, or build a useful cabinet, or ski smoothly down a steep powder slope. When you are struggling in some particular area, it is easy to raise its importance and worth. If you already fear mathematics, you don't need the additional obstacle of thinking that it's something other than a useful tool.

It is also important to put mathematicians on the same level as other people. Students believe math teachers can add long columns of numbers in their heads or get the right answer to every problem immediately, with little effort. When you believe the myth that math teachers are smarter or quicker than the average person, you create serious problems for yourself. Remember the math instructor has studied math seriously for an extended period of time. If you had been doing these same problems for as long as he or she had, you would be good at them too. So no comparisons!

THE IT-SHOULD-BE-EASY MYTH

This brings us to the common misconception that those who do well in mathematics always find it easy. Teachers reinforce this as they write problem solutions on the blackboard. They are usually done right the first time and presented in finished, logical form. We think we should be able to solve problems with similar ease. Rarely does an instructor inform us of how long it took to do a similar problem the first time or how many false starts it took on this one. We think something is wrong with us when things seem to take forever.

Finding the answers to difficult math problems is not terribly easy, and inspiration does not come quickly. Repetetive or drill type problems sometimes go fast, but they are the exception. Besides, drill problems are only the necessary mathematical drudgery. It is like becoming fit before running in a 10 kilometer race. The race may be fun but getting in shape beforehand is hard work. Drill problems in mathematics are necessary before being able to understand other more interesting aspects. But being able to add up long columns of figures quickly is not what mathematics is about. In fact many mathematicians admit they have trouble doing routine mathematics, like balancing checkbooks.

Many mathematicians readily admit they have trouble doing routine mathematics...

It is important to realize that having difficulty in solving problems is not unusual. An instructor's clear logical presentation on the blackboard is a result of long experience and advance preparation. You should expect to stumble at first in solving similar problems and come to full understanding only in slow stages. You will find that problems that are the most satisfying and exciting to solve are those that you've had to struggle with for awhile. Easy problems, like easy games that you always win, aren't much fun.

THE UNIQUENESS MYTH

When you're in what you think is a desperate situation, it's easy to believe that you are all alone, that no one else could possibly feel what you are feeling or fail where you are failing. It's embarrassing to admit you're uncomfortable doing simple mathematics so you don't tell anyone. If you dread having to compute the tip to leave at the end of your restaurant meal, it is a normal response to believe that no one else feels quite so uncomfortable.

When people are in a setting where they feel free to admit their mathematical weaknesses and terrors, it is amazing how many have similar feelings. In a math class when some brave individual finally asks a question or says they don't understand, you can almost hear the sigh of relief from the rest of the class. Yet most of us continue to struggle with mathematics, and our negative reactions to it, alone. We are afraid to admit we don't know for fear everyone will think we're stupid. Also we don't ask for help with problems because we've been led to believe that math is something you should be able to work out on your own. Try to remember the last time you had an exciting discussion in your math class, where people worked together solving problems, making suggestions, and sharing feelings they were having in the process. Unfortunately you probably can't remember many and that's too bad. Such discussions might help to convince us that our questions and frustrations are common ones, shared by most people who attempt math.

THE GENDER MYTH

It has been popular to explain the absence of many women in advanced mathematics with the theory that women are just not as able or as interested in this field as men. It is unpopular these days to single out any particular group and claim they are inferior in some way. Yet this does not mean that in some circles these beliefs are not alive and well. In fact, in 1982 we still hear the story of a Dean of Engineering at a major university telling a young woman applicant that women do not make good electrical engineers—there is too much math and logic involved. He ended up suggesting that she consider some other branch of engineering instead. Of course 20 years ago he probably would have suggested she not consider engineering at all. It is interesting to note here that when telephone switchboards and typewriters were first developed, women were considered unable to perform the complex operations necessary to run these machines.

Even though some people prefer to believe otherwise, none of the studies done in the area of gender and math performance have established that one sex or the other is better suited to do well in math. What has been shown is that women generally score lower on achievement tests after about the eleventh or twelfth grade. Most of these studies fail to take into account that by that time the majority of women have taken less math than most men have. Women also score lower than men in some tests of spatial skills* and problem solving ability.[†] No one, however, has been able to demonstrate that these results are due to being female or male. Cultural experience and development seem at least as significant. Until women and men have the opportunities to be involved in the same types of activities throughout childhood and early maturity, it will be impossible to determine the causes of differences in math performance.

There are many good articles listed in Appendix C that explore this question in detail. The important issue for us is not why women and men on the average perform differently in math. Instead it is how individual women and men can reach the level of mathematical proficiency that they need or desire. If you are a man or a woman struggling in your efforts to learn mathematics, the information that men score a few percentage points higher on standardized math tests is not helpful.

There has also been the temptation to label math anxiety as mainly a woman's problem. Much of the literature available has been written from the feminist perspective. In an effort to understand reasons why few women were entering technical and scientific fields, researchers noticed that women chose to take less mathematics than men during high school and college. Women's attitudes about math were sampled thoroughly but men's were not. Later studies have shown that negative reactions to mathematics are not limited to females. Although women may more readily admit feelings of anxiety and fear, men are now expressing similar reactions to math. Pressure from parents and counselors may lead men to remain in math courses longer than women, but they often experience the same intense dislike of the subject. When given the opportunity to deal with their emotions in math anxiety reduction programs, men have turned out in surprising numbers.[‡]

---

*These are tests that measure the ability to visualize objects and their movement in space. For examples, see Chapter 8.

[†]There is one group of women whose spatial skills are comparable to the scores of men; that group is women athletes. There are also some cultures where men and women have no measurable difference in spatial skills, Eskimos for one.

[‡]At one point in the program at Wesleyan there were 40 women and 30 men participating in the program on math anxiety.

THE RIGID, UNCREATIVE MYTH

Some people become turned off to mathematics because they view it as a dreary, overly structured body of knowledge. It seems to have no history, no relevance, and little future. They see it as unchanging, logical, inflexible and uncreative. This belief is confirmed when they learn that math is organized into separate and seemingly unrelated blocks like arithmetic, algebra, geometry, calculus, statistics, etc. Unfortunately, mathematics teaching in school often reinforces these notions in several ways.

First: The history of mathematics, together with its fascinating characters, is rarely discused. For example we never hear about a mathematician like Leonhard Euler (1707–1783) whose life was remarkable as were his contributions to mathematics. He is a contrast to the image of the mathematical loner working in a small stuffy office creating theorems*. Euler authored 20 volumes of mathematical works, doing much of his work at home among his 13 children. The story goes that he used to write a chapter between first and second calls to supper with a child sitting on each knee. He continued to do mathematics until the time of his death even though he was almost totally blind in his later years. He is just one example of the many interesting individuals who have contributed creatively to mathematics.

Second: In school, we are taught that there is a right way and a wrong way to do every problem. The "right way" is the method of the textbook or the teacher or the one that has been around the longest. The "wrong way" is anything else. Imaginative and nonstandard solutions proposed by students are rarely accepted. Evariste Galois (1811–1832) was a rebellious French mathematician who made dramatic advances in several branches of mathematics although he died in a duel at the early age of 21. He once said, "a student of superior ability is wasted on an examiner of inferior ability." Students can become discouraged when their unique methods or approaches are not considered the "right" or "proper" ones. Galois, for example, finally left school in disgust.

Finally, intuition is often discouraged in the classroom. Math "intuition" means knowing an answer without knowing exactly how you got it. Intuition is unacceptable despite the fact that it is the *cornerstone* of mathematical discovery. We are led to believe that if we can't explain it logically, then it isn't right. But logical explanations are really one of the end products of mathematics. Teachers forget that

---

*Theorems are just statements of mathematical facts. Discovering and proving them is largely the creative work of mathematicians.

all *new* mathematical discoveries come about through the use of intuition and intelligent guessing. Students are told that hunches should play no part in learning mathematics.

A large poster seen in a junior high school classroom says

THOU SHALL NOT GUESS.

This notion of not guessing and ignoring intuition leads to dire consequences. We come to distrust even our good common sense. The poster mentioned above might just as well say "THOU SHALL NEVER WORRY ABOUT THE ANSWER BEING REASONABLE." Although few mathematics teachers would suggest that this rule was desirable, it is one of the outcomes of discouraging intuition. To a question "How many people attended the show?" students will turn in answers like $21\frac{1}{2}$. They will rarely notice that a half a person doesn't make sense. To a messy division problem like "find $36\frac{3}{7} \div 2.934$," answers like 84 will be proposed after a long string of manipulations. Surely this can't be correct. Can you see why?

Try to make a guess at a reasonable result. If you said $36\frac{3}{7}$ is close to 36 and 2.934 is close to 3 and then divided these two values, you might see that the result should be somewhere around $36 \div 3 = 12$. This is very different from 84. In math classes we are rarely asked to make reasonable guesses. Consequently, we often come up with ridiculous answers, preferring to trust a rule or formula we don't understand, rather than our own intuition. No wonder we think math makes no sense.

THE USING-TOOLS-IS-CHEATING MYTH

When was the last time you confessed that you count on your fingers, or use them when you're doing arithmetic? Most of us feel slightly embarrassed if we are caught in the act. There is a mistaken view that to use aids in doing math is the same as cheating. Is it cheating to come up with creative methods to make tasks simpler? Until recently, the use of calculators to do basic arithmetic was also discouraged. Is it cheating to eliminate the drudgery of mathematics and to improve accuracy? To answer "yes" to the last question is like insisting that we all go back to a prior age and make our own butter from scratch, beginning with raw cream and a churn. After all, the argument goes, how can you really say you understand butter if you buy it ready-made at the supermarket. It has been said that if we use calculators to do arithmetic we will never really understand how to do it without them. That case has been overstated.

It is probably true that we will not be as skillful at adding long columns of numbers quickly. So what? In order to use a calculator effectively, it is still necessary to understand many of the mathematical principles involved in doing the operations longhand. This is the significant information we need from arithmetic. To do complicated operations longhand has little value in itself. Today we have the advantage of the calculator—a time-saving, accurate tool to help with the details of arithmetic. Without this tool the details are often tedious and boring. To avoid its use is not reasonable. It is probably more appropriate at this point to insist that people learn how to use a calculator effectively than it is to insist that they memorize multiplication tables. (It is still important to understand how multiplication works however.) We should be open to using any tools that are available (fingers, toes, calculators, etc.) to make our mathematical lives more enjoyable. It is the decision whether to add or multiply in a particular situation that is the real mathematics, not the carrying out of the operation.

THE NO-ONE-BUT-ENGINEERS-USE-IT MYTH

Many people feel they don't need to take math. It seems to them that science, mathematics, and engineering are the only fields that require math skills. They believe that most people can get along in everyday situations without it.

Chapter 4 discusses many reasons why it is important to take math. At this stage, we will only say that most people need to be able to do simple math daily. Counting change at the checkstand, paying bills, figuring out interest on a loan, reading newspaper articles with statistics and graphs in them are examples of everyday uses. In addition many jobs—not just those in the sciences—now require math. Saying you won't ever need to use math is an excuse that has serious negative outcomes as you try to earn a living and to survive in a very technical society.

CONCLUSIONS

You may have recognized some of the ideas described above as part of your own belief system about mathematics. The authors believe that those who support very many of them will have problems functioning well mathematically.

If you believe

- there is such a thing as a "math mind," you don't have one, and there is nothing you can do about it;

- mathematicians are geniuses;

- math should be easy;

- you are the only one that feels this incompetent;

- women are usually poor in math (if you are a woman);

- math is rigid and boring; or

- you will never need to know math

then you will either avoid math entirely or become discouraged too quickly when you meet obstacles. Developing a more positive attitude about these matters is an important step in putting math back in its proper place.

FURTHER READING

Bell, Eric. *Men of Mathematics*. New York: Simon & Schuster, 1937.

Osen, Lynn. *Women in Mathematics*. Boston: MIT Press, 1974.

Resnikoff, H.L. and R.O. Wells, Jr. *Mathematics in Civilization*. New York: Holt, Rinehart, & Winston, 1973.

and how to approach math effectively

3

WHAT IS MATH?

This chapter will describe characteristics of math that may make it seem difficult such as:

- *it uses symbols,*
- *it uses words in a special way,*
- *it requires some memorizing,*
- *it demands active involvement,*
- *it involves the use of logic, and*
- *creativity and other factors play a role.*

In this chapter we begin to talk about the subject of mathematics itself. As you read about it, you may start to feel some of the same anxiety or apprehension that you experience when actually doing math. Now is the time to acknowledge any of those negative feelings but not to let them stop you. Don't feel that you must thoroughly understand the symbols or math topics that are discussed. They are used only as examples to help illustrate a point. We hope that knowing about some of the special qualities of math may make it less frightening and easier for you to learn.

MATH USES SYMBOLS

Mathematical facts are usually expressed in shortened form. For example $D = R \times T$, means the distance traveled is the rate of speed multiplied by time. For example, if I drive for 2 hours at 50 miles per hour, the distance I travel is 50×2 or 100 miles. And $A = L \times W$ means the area (in this case for a rectangle) is found by multiplying length by width. There is often not a great deal of explanation along with these symbolic statements. In a basic math textbook you may see pages and pages of symbols and shortened statements. Ironically, much more advanced math books often look less frightening because they seem to have more conventional sentences and paragraphs. The advantages of a mathematical shorthand are well known, however. An entire idea can be expressed in a single symbol. You have less to write and you can be very precise. For example, the symbol "$\leq$" means "is less than or equal to." Having a symbol for this important concept reduces the length of statements that contain it.

Yet when you see a page of symbols it may make you feel like you are trying to understand a complicated idea written in a foreign language. At first glance, the problem of translation seems overwhelming. The statement "$2b \vee \sim 2b$" is one you recognize, or would recognize if only you knew the meaning of $\vee$ and $\sim$. In logic (and in mathematics), $\vee$ stands for the word "or" and $\sim$ means "not." Now read again: "two b or not two b." Is that a familiar phrase? The statement is a mystery only until you learn the meaning of a few unfamiliar symbols.

Symbols must be written and read very carefully. This also causes problems for the person who is distressed by math. Two different symbols may have very similar meanings, but are different in important ways. And two symbols that we may use interchangeably in everyday life, like "E" and "e", are considered different in math.

Usually there is only one symbol that is correct in a given mathematical statement. Thinking of new ways to express an idea may be

helpful in English composition but creates some bad habits for us in math. So learning to read and write symbolic mathematical statements is like learning a new language, but a new language with additional complications. Not only must we learn the meaning of new words and symbols, we must also learn to be much more precise than may be our habit.

MATH USES WORDS IN SPECIAL WAYS

The situation gets worse. We have seen that math uses symbols, which often have no meaning for us outside of this subject. In addition things are made harder by the use of everyday terms to describe very specific concepts in math. Sometimes the mathematical meaning is close to common usage and sometimes not. In either case we may have problems understanding. Consider the word "field" for example. Take a moment and decide what meaning the word has for you. When mathematicians discuss a "field," they are thinking of a collection of objects (like numbers) with two operations defined on them (like addition and multiplication) that must satisfy about a dozen different conditions. Perhaps you can see the potential problem. The non-mathematician hears "field" and thinks of a green, grassy enclosure, while the mathematician hears the same word and is thinking something quite different. In this case the special meaning assigned to the word in math doesn't resemble its everyday usage at all.

When words have meanings that are quite close in math and English, the problems are often worse. The word "multiplication" is a good example. According to the dictionary to multiply is "to increase in number; to add quantity to . . ."*. But then the results of multiplying fractions may seem unreasonable. When you multiply two fractions like 1/2 and 1/3, you get 1/6, which is smaller than either original number. To "multiply" in mathematics is to perform a very specific process that obeys a set of laws. For whole numbers (like 2 multiplied by 6 gives 12) the result fits with our notion of getting larger. But these same laws, applied to multiplying two fractions, go against our idea of what it means to multiply. In order to do math well, we must accept specialized mathematical meanings and be willing to expand some of our old comfortable beliefs.

Sometimes we don't realize how much of our everyday language consists of words and concepts that have mathematical meaning. The very fact that we understand ordinary conversations means we must actually know quite a bit of mathematical vocabulary. The exercise that follows shows how extensively we use mathematical language in our lives.

---

*Webster's New Collegiate Dictionary, Springfield, Mass.: G.&C. Merriam Company, 1975.

Exercise: MATH VOCABULARY

Directions: In your non-mathematical reading in the next few days see how many of these words (or their variations) you can find. In addition write a non-mathematical sentence containing each.

add (addition) _____

big (bigger) _____

contradiction _____

divide (division) _____

equal (equality) _____

equivalance _____

false _____

function _____

game _____

if (some statement), then (something) _____

_____

larger _____

less _____

multiply _____

necessary _____

negative _____

only if _____

probable (probability) _____

_____

problem _____

prove _____

sequence _____

series _____

statistic(s) _____

subtract (subtraction) _____

_____

sufficient _____

true (truth) _____

valid _____

MATH REQUIRES MEMORY

The use of language and symbols in mathematics has just been considered. We must learn and remember the special meanings of words and symbols within the subject in order to proceed. This requires very careful, precise memorizing. The role of memory in mathematics, however, is often misunderstood. Students sometimes complain that math consists of unrelated rules that make no sense; the only choice is to memorize them. Nothing could be further from the truth and nothing leads more surely to failure as trying to understand math by memorizing it.

It may be hard to believe at first, but math does not consist of unrelated facts and formulas. When you take the extra time to look at all the pieces together, you begin to see the relationships and connections. It is always an exciting time when that happens and it makes further study of mathematics easier. It is a sign of real progress when you first encounter a formula that finally makes sense to you. You can figure it out from other things you know and it no longer needs to be "memorized." Other connections will come faster because you now believe they exist. This should be reassuring if in the past you've been overwhelmed by vast amounts of material that must be learned. You were right to feel distressed. You cannot experience continued success in mathematics if you try to get through by memory alone.

MATH DEMANDS ACTIVE INVOLVEMENT

Now that we've said you needn't spend all your time memorizing formulas and operations, you may be wondering what is left. Lots! Ask any successful mathematics student and he or she will tell you that math is not a spectator sport. By that we mean that you cannot learn to do mathematics by watching other people do it. People who learn to play a musical instrument have to practice regularly. In the same way, someone who is learning to do math will not come to understand concepts by passively reading textbooks, or listening to math lectures, or by watching other people solve problems. These things may be helpful to get you started. But you become skilled only by doing lots of problems yourself. The shakier you are on a topic, the more problems, of greater variety, you should attempt. In addition you must always test your own understanding by asking why you may do such and such a step, what problems are similar, what ones are different and why. What generalizations can you make? It is not sufficient to merely follow the steps on a sample solution. Your involvement in the process must be active.

MATH INVOLVES USING LOGIC

A discussion of the nature of mathematics would not be complete without mentioning the role of logic. You can understand mathematics only when you appreciate the orderly relationship between definitions, assumptions, and the conclusions based on these. These relationships are the very basis of a mathematical structure, and the language of formal logic is the vehicle for conveying them. Yet there is a real difference between learning to recognize logical connections and organizing our daily life in a "logical" manner.

Comments abound of the following sort: "I could never be good in math; I'm just too illogical." We suspect that the illogical reference here probably relates more to the way a person's life is ordered than to their ability to follow a reasonable argument. It is the latter that is relevant in mathematics. Although there may be some connection between orderly thinking while solving math problems and orderly behavior and reasoning in the rest of your life, the connection is by no means direct. Some very good mathematicians have personal lives that seem extremely chaotic. At the same time, there are people with very orderly lives who experience great difficulty doing math. It is our opinion that almost everyone has a sufficiently developed intuitive sense of logic to study math successfully. We sometimes fail to recognize that we use logic and sequential reasoning in the tasks we perform routinely in our daily lives. For example purchasing the ingredients for an elaborate dinner and preparing the dishes so that they are ready at the proper time, involves an extensive amount of orderly thinking and planning.

THE ROLE OF CREATIVITY AND CONFIDENCE

Solving a challenging mathematics problem is the result of complex processes taking place in the right order. Your success often depends on something we call mathematical maturity. The more similar problems you have seen and the building blocks you have learned, the more mathematically mature you are. Success also depends on your working atmosphere, both external and internal—can you concentrate or are you distracted by outside noise, confusion, other more pressing thoughts or personal problems? Your confidence level is significant—do you believe you can get the answer or do you worry that it's too difficult for you? Studies have shown that people's success in solving problems depends heavily on their own expectations. In one experiment, a group of people is given a problem and told it is quite simple, that everyone with a little effort should be able to find the answer. A second group of similar ability is given the same problem but told it is very difficult and that only a few people will be expected to get the right result. The success rate in the group given the "easy" problem can be two or three times as great as the rate in the group that thinks the problem is "hard."

What about your level of trust? Sheila Tobias says, "People who don't trust math may be too wary of math to take risks."* Are you willing to risk being wrong or being unsure? Do you believe the problem is fair or are you really wondering if there is some trick involved? Will you agree to accept the information, definitions, etc. that are given or do you doubt their validity? There is certainly a place for healthy skepticism in mathematics but you also need to get into the spirit of the problem. Are you really trying to solve it, or are you getting angry and looking for a way out?

Finally, the speed with which you solve a math problem involves an element of luck. Often a spark of insight provides the key to the solution. What makes the light bulb glow, the "aha" reaction occur, is not really understood by either science or psychology. Hearing the right word or seeing an object that helps to make the connection is often as much a matter of good fortune as it is of careful thought. We will have more to say about problem solving later. The important point here is to notice that successful solutions involve complex processes, which can happen slowly or quickly depending both on circumstances and mathematical experience. This is true for the expert in mathematics as well as for the beginning algebra student.

---

*Tobias, Sheila. *Overcoming Math Anxiety*. New York: W.W. Norton and Company, Inc. 1978, p. 53.

how can it possibly help me?

4

WHY STUDY MATH?

This chapter describes five reasons for taking math. These are
- *to keep career options open,*
- *to give advantages in test taking,*
- *to be a careful consumer,*
- *to keep up with future trends, and*
- *to get personal satisfaction.*

In deciding if we want to overcome math anxiety, it's important to be realistic. Do your personal goals require that you take more math? If math is a difficult subject, then studying more of it is going to require a real commitment of time and energy. In addition, for those who experience math anxiety, taking more math may at first bring up feelings you'd rather avoid. Here is some information that we believe will help you decide that learning more math is a worthwhile investment.

KEEPING CAREER OPTIONS OPEN

Many adults find that decisions made as teenagers, regarding how much math they would need, are not now realistic. Often these decisions were made when entering high school. The less math the better was often the deciding factor. But, in spite of what was thought at the time, people's situations change. Needs and requirements do also.

Whether you are now just entering high school or are an adult returning to school to change careers or advance in your present one, you need to ask some difficult questions. At this moment, what career do you think you might like to pursue? What are the educational requirements for that job now and what might they be in the future? Are the educational requirements different for advancement in the field than they are for entry? We have a few words to say about each of these issues.

First, some more questions about your current goals and ambitions. How permanent do you think your choice about careers will be? Do you know now what job you will want to be doing 10 years from now? Most people decide on a career direction with limited knowledge about what it involves and many change their minds when they actually enter the field. In addition, changes in technology constantly make some jobs obsolete. Much work once done by people is now performed by machines. Workers are often forced to retrain in a new area where job opportunities still exist. In spite of these facts, most people prepare themselves for only one career choice, and shut themselves off to other options. Yet we know that the majority of people who work will change type of employment at least once by choice or necessity. It therefore makes sense to acquire an educational background that will prepare you for as many different areas as possible. There is probably no subject as universally required as mathematics.

So what are the mathematics requirements for various fields? There is a widespread belief that mathematics is needed only for a few technical fields like engineering or statistics. For one thing, we all know very successful people who have taken little if any formal math training and have done just fine in a variety of jobs. Yet the basic requirements for entry into many fields are stricter than before. There is now more intense competition for good jobs and to get them you must be prepared. At the end of the chapter, you will find the math requirements listed for a variety of majors at a typical four-year university. Even if not much math is done in the performance of a job, there may be a significant math requirement *just to get in* the program of study. It is estimated that 70% of the college majors need a full four years of high school math and at least one college level math course. It is interesting to note also that those areas that demand the greatest amount of math are those that currently have the most open jobs. Look at the "positions available" column in your local newspaper if you have any doubts.

A significant amount of mathematics is needed in vocational and trade areas as well. Many of the tests for admission to apprenticeship programs in such trades as carpentry, plumbing, and electrical work consist almost entirely of math. Without an ability to do at least algebra and geometry, the chances of even being admitted to such programs is slight.

Fields that do not require math to enter are almost certain to demand some to move up. Supervisory jobs in most fields require knowledge of budgeting and the use of statistical information. You must also be able to apply standard formulas used in the field.

Those who don't have these skills and who resist learning them, are usually the ones stuck in the lowest level jobs in any field. Also even if a mathematical background is not a stated requirement, having some special skills in this area may provide just the edge in getting a job. For example, jobs in anthropology (which requires no math) are quite scarce. But anthropologists who also know computer programming are much in demand.

TEST-TAKING ADVANTAGES

Knowing basic mathematics is helpful in doing well on most of the tests required for admission to college. The College Entrance Examination Boards, the American College Tests, the Graduate Record Exam, and most of the professional placement exams all have sections dealing with math skills. It has also been demonstrated that taking algebra and geometry gives a 25% advantage on civil service exams. These are the tests that are used to screen for jobs with the city, county, state, and federal governments. Having taken just high school algebra can mean the difference between qualifying or not. It can also affect the level at which you enter and the salary you will earn.

Remember, fields that require very little math in practice, use math as a placement tool. Many potential employers use math scores as a measure of how readily you can learn a new skill and be trained for their jobs. If for no other reason, it may be helpful to acquire additional math skills just to improve your general placement test scores.

AWARENESS OF CONSUMER ISSUES

Being able to do some basic math is essential in everyday tasks. These include balancing checkbooks, counting correct change, figuring out tips in restaurants, verifying discount prices, reading tables and graphs in newspapers, and passing judgment on the reliability of the statistics we read. One study has shown that 60% of Americans cannot use basic math to work such common tasks. This costs them over $100 million a year in overcharges and losses. There are those people who we all must sometimes deal with financially, who count on us being unable or unwilling to check facts and figures.

It is also a mistake to think that if you need some math work done you can easily hire an "expert" to do it. Discomfort with math makes it difficult to hire specialists and to watch over them properly. Certainly there are some parts of mathematics that you might not want to master yourself. But in order to be an intelligent purchaser of math services you need to be willing and able to ask the right questions and to give reasonable directions. For example, if you take out an auto loan, you may not need to know how to calculate the various interest rates and installment payments. But you should be able to figure out where you can get the best deal. If you are embarrassed by your lack of math ability, you may be unwilling to ask the proper questions, and end up paying much more than you should. Another example is the use of computer consultants. To many of us, what goes on inside a computer is so mysterious that we are in awe of anyone who understands them. But in order to make efficient use of computer experts, the person with the problem to be solved must be directive and demanding. It also saves time and money if they have some knowledge about what computers can and cannot do.

KEEPING UP WITH TRENDS IN TECHNOLOGY

We have already said that there is an increasing rather than a decreasing demand for skills in math. We expect that in 10 to 20 years, computer terminals will be as common in homes as TV sets or telephones are today. Understanding how computers work will be essential to doing everyday business, to keep track of data, and help make decisions. If you cannot or will not use them you will be at a serious disadvantage.

As the information explosion continues, the use of statistics and computer technology to manage that information will expand. The average business person will not be able to make informed decisions unless they understand some basic statistics,* can read graphs intelligently, and can use probability theory.† If this seems farfetched, think about how things have changed in the last 20 years. Who would have guessed that home computers and programable microwave ovens would be commonplace items today? Mathematical modeling, game theory, and decision theory, will soon be household words for most people, not just the specialists.

---

*Statistics is a branch of mathematics that deals with how to manage data and information.

†Probability theory is the systematic study of how likely it is for some event to happen.

GETTING PERSONAL SATISFACTION

Perhaps the best reason to take on a difficult task is because you want to, not because you have to. Many people who have avoided math for lots of years decide they need to study the subject further just for themselves. They want to experience the feeling of success that comes from doing well in an area that has until now frustrated them. Often they are now a very different person from the one that gave up on math sometime in the past. They recognize that they have suffered from math anxiety or avoidance and want to do something about it. They are ready to take back control over this part of their lives.

People who have reached this stage often find that studying math now is much easier and less frightening than they remember. It still involves hard work and commitment to the task. But it is not overwhelming or even painful. And the confidence and satisfaction they feel when they conquer math hurdles increases their sense of self-worth. Many report that they become more confident, not only in math related situations, but in other areas as well. Not only do they use calculators in public without apology, they may learn how to tune cars. They may also try to operate a nonautomatic camera, or take up scuba diving, or rewire a lamp. It appears that when we question our ability to do mathematics, we also don't trust ourselves to do well in other areas dealing with technology and mechanics. Overcoming math blocks can be a truly liberating experience.

ACADEMIC REQUIREMENTS

The chart on the following page may help to illustrate how many fields of endeavor require at least minimal mathematical competency.

FURTHER READING

Remick, Helen. Participation rates in high school mathematics and science courses. *The Physics Teacher* May, 1978.

Sells, Lucy. High school mathematics as the critical filter in the job market. In Developing Opportunities for Minorities in Graduate Education, Proceedings of the Conference on Minority Graduate Education at the University of California, Berkeley, May 1973, pp. 47-59.

SELECTED MAJORS AND THEIR REQUIREMENTS*		
College major	Years of high school math	Quarters of college math†
Agriculture	3	3†
Animal science	4	2†
Architecture	4	3
Biology	4	5
Business administration	4	3
Chemistry	4	5†
Child and family studies	1	1
Computer science	4	5†
Construction management	4	2
Criminal justice	3	1
Education (Elementary)	2	2
Engineering (General)	4	6†
Environmental science	4	3
Forestry	4	3
Geology	4	3
Horticulture	4	2
Hotel administration	3	3
Interior design	2	0
Nursing	2	1†
Pre-medicine	4	1
Psychology	4	4
Recreation	3	1
Social work	1	1
Zoology	3	2†

*Taken from "Don't Be A Math Dropout," Washington State University, Pullman, WA. 1980.

†Number of college courses required depends on several factors: high school background, particular college requirements, and specialty within the major.

learning what is in my control

5

SELF-
MANAGEMENT
SKILLS

This chapter will discuss several personal skills that can help you change your behavior (and your feelings) toward math. These are techniques for
- *eliminating self-defeating behaviors,*
- *communicating effectively, and*
- *reducing stress.*

After reading the first four chapters and thinking about your math history, you are more aware of the causes of math anxiety and of the mistaken ideas we carry around with us about math. You also probably understand that there are features of math that make it different from other subjects you have studied. One purpose for this background information is to help you get a perspective on math and to recognize that you are not alone. You may find that you hold some biases about math or about yourself that you want to set aside, at least temporarily.

Even so, you probably believe that any future math experiences you choose to experience will still be largely out of your control. After all, you can't tell instructors how to teach and you can't make yourself smarter, so what could you possibly change? The answer is, you can take charge of your own behavior. You can learn new skills that allow you to control a great deal more of your math encounters than you ever imagined. We are calling these skills *self-management skills* and defining them as skills that you use to modify and change your behavior. They include

1. skills for acting calmly under stress,
2. skills for strengthening your positive beliefs about yourself and your abilities, and
3. skills for making your ideas, feelings, and needs known to others.

In self-help books, these are often described as

1. skills for eliminating self-defeating behaviors,
2. assertive communication skills, and
3. stress-reduction skills.

You will find exercises in this chapter to assist in identifying specific skill areas and give you practice in improving these skills. Not all of these approaches will work for you. Try each of them, then choose the ones that fit best with your personal style.

SELF-DEFEATING BEHAVIORS

Exercise: SELF-DEFEATING BEHAVIORS

Directions: Read through this list and circle any items that could describe your behavior in the situation described.

1. I always have trouble finding a parking place, so I arrive in math class 10 to 15 minutes late.

2. I have to go to school full-time and work full-time, which makes it rough to find enough time to study math.

3. It's hard for me to get up in the morning and I'm taking an eight o'clock class; I just can't seem to get there more than half the time.

4. I usually put off studying until the night before the math test; that way I only have to be nervous for one night instead of several nights.

5. I never ask math instructors for help. That way they don't know how little I understand.

6. I spend a lot of energy worrying about how I'm going to do in math.

7. I don't have time to do homework every night but I try to catch up on the weekends.

Any one of these behaviors can be self-defeating. That is, by doing them you can prevent yourself from reaching a goal that's important to you.

Coming late to class, skipping class, not doing homework on time are all self-defeating and are generally activities within your control. You might wonder why you keep doing them if they're things you don't have to do and they are hurting you. The problem is that we don't always act in our own best interest. We don't always eat the right foods even though we know what they are. We don't always get enough exercise despite the fact that it makes us feel good. Sometimes we have done something for so long, putting off homework, for example, that it becomes a habit, just like overeating or smoking become habits. Some habits are harder than others to change. But habits, including those that are self-defeating, *can be changed.*

If you do decide you want to change a self-defeating behavior, here are some steps that will be helpful:

1. First become aware of your behavior. Pay attention to what you are doing for a week. Ask yourself whether you have done any-thing that might be hurting your performance in math (or in other situations in case you are not enrolled in a math course). This is not a time to judge yourself or to make resolutions to change completely. It is a time to collect data on yourself. This is the time to answer the question "Are any of the things I do working against me?"

Sometimes we have done something for so long, putting off homework, for example, that it becomes a habit.

2. Once you have identified your self-defeating behaviors, choose *one small one* you would like to change or modify. Trying to change everything at once just won't work. That becomes a self-defeating behavior in itself. You are setting yourself up to fail if you try to eliminate at once, habits or behaviors that may have taken years to acquire. You will end up convincing yourself that change really is impossible.

3. Now put what you are going to do down in writing. You will have your goal stated clearly and openly. You'll also be less tempted to revise or modify it later. This is called a contract for change and an example is given on the following page.

Many people find it lends more power to the contract to have an instructor, counselor, classmate, or friend co-sign it. You may use the blank contract that follows (or some format of your own) to write down your own plan for change.

In addition to using self-defeating behaviors, we also tend to use a wide range of self-defeating thoughts.

Exercise: SELF-DEFEATING THOUGHTS

Directions: Read the list below and circle thoughts you feel you may have had at some time.

1. Some people have what it takes to do math; I don't.

2. I'm good at some subjects, but not math.

3. I don't want to take math because I know it will hurt my grade point average.

Behavior to change	Steps to change	Target date	Resources needed	Reward
Example: I'm going to get to class every day next week.	1. Set alarm ten minutes earlier. 2. Warn kids we are getting up earlier next week. 3. Make lunches at night. 4. Go straight to available parking area instead of driving around hoping to get lucky and find a closer space.	Start Monday the 4th and finish Friday the 8th.	1. Call Helen and ask her to remind me to make lunches at night 2. Ask the kids for their cooperation; tell them its impor- to me.	Hopefully, understanding what the teacher is talking about will be reward enough. If it isn't, on Friday I'll have lunch with Janet.

I agree to work on this plan for change for _____.
(period of time)

I will call on other people to help me if I think that will help me carry out this plan.

_____   _____
(signature)   (date)

Behavior to change	Steps to change	Target date	Resources needed	Reward

I agree to work on this plan for change for _____.
(period of time)

I will call on other people to help me if I think that will help me carry out this plan.

_____   _____
(signature)   (date)

4. My math teacher is really boring and can't explain things at all! I could do math if it weren't for him/her.

5. I'm a real failure when it comes to math.

6. I can do calculations but when it comes to word problems, forget it. I just can't do those.

7. My brother (sister, mother, father) always made me feel dumb, so now I can't do math.

8. If I ask someone for help, they'll think I'm stupid.

These thoughts are examples of either blaming someone else for your math problems or putting yourself down. Both pointing the finger at someone else and feeling bad about yourself have the effect of reducing your ability to do math. Therefore, they are self-defeating thoughts.

The difficulty with self-defeating thoughts, is that they come to us automatically. You may be so used to defining or picturing yourself as "poor in math" that this *assumption* becomes a truth in your mind. It then becomes the only way you view yourself. It may be true that for various reasons—some beyond your control—you had difficulty in math classes starting very early in your life. You may be distorting that experience, however, when you carry it into adulthood and use it to define yourself as "stupid in math." It will require a special effort to change your self definition and to look at yourself in a new light. After all, for years you may have been telling yourself you are a math failure. How could you believe otherwise?

Picturing yourself as stupid may become the only way to view yourself.

One tactic for change is to become very conscious of your thoughts. When you are struggling with a math problem or trying to figure out a tip in a restaurant, ask yourself, "What am I thinking now? What am I telling myself about me and math?" Your first temptation may be to say, "I'm not telling myself anything. I'm too nervous to think about anything." You are probably mistaken. If you are very attentive to what is going on in your mind, you will be able to recognize some of your self-defeating thoughts.

The next step is to replace these negative thoughts with positive ones about your math abilities or experiences. In time this will help you redefine yourself with respect to math. This technique has been used successfully by many athletes and business people. These are people who cannot afford to have their throwing arm be "off" or their sales declining. They train themselves to examine how they are thinking about their performance. They know that worry and doubts about their abilities have the effect of reducing those abilities. When they replace negative thoughts with positive ones they interrupt the self-defeating cycle that has been established. Here are examples of some positive statements you could say about yourself.

- I am a capable student and therefore I am also good in math.

- I can approach each math problem with an open mind.

- I am a curious person and I like to figure out how things work and to solve problems.

- I see math as a challenge for me, not a hopeless problem.

These statements do not have to be true at the time you say them. On the other hand, they are not wishes about how you would like to be or New Year's Resolutions; they are merely positive, descriptive statements that may be just as accurate in portraying you as are the negative ones that have become a habit.

Exercise: POSITIVE STATEMENTS
Directions: Think of some other positive statements that could describe you and list them here. Decide on at least four.

1. _____

2. _____

3. _____

4. _____

Now find a 3" X 5" card and write the statements that you think fit you best. If you are taking a math course, carry the card with you and read it several times a day. Read it when you wake up, when you enter your math class, when you leave your math class, and when you go to bed. If you are not in a math class, tape the card to your bathroom mirror and read it several times a day. After you have read your statements, try to close your eyes briefly and imagine yourself as this person who has a positive relationship with math. It may take several months before these positive thoughts become as automatic as the former, negative ones. Stay at the exercise until they do.

It may take some effort before positive thoughts become as automatic as negative ones.

ASSERTIVENESS

Assertiveness is another self-management skill. By assertiveness we mean the ability to communicate directly to others such things as our goals and plans, feelings, needs, and ideas. Often the term "assertiveness" causes confusion or even negative feelings. To avoid this possibility, assertiveness here will refer only to a set of communication skills. These include the skill of asking questions, of saying "no," of giving your opinion, and of asking for help when you need it. The choice of whether or not to use these skills is always yours. But in order to have this choice available, you must develop the skills; otherwise you limit yourself to only one way of doing things.

First we need to answer the question, "What do communication skills have to do with mathematics?" Many of your math encounters occur in a social context. They occur in a math classroom; with a tutor, friend, or relative; or with a math teacher. You need to be able to communicate with these people. In addition, if your math encounters are often stressful, you may need to communicate your feelings and concerns to others.

Here are two examples of situations where good communication skills would have helped a math student.

- *Marlene completed her math assignment and compared her answers to those in the back of the textbook. On one problem the answer was different from the one she gave. She didn't understand this answer but thought hers was correct. She really wanted to ask the instructor about this, but the instructor didn't appear to be a patient person and Marlene didn't want to take a chance on exposing herself to discomfort or humiliation. She said nothing and just changed her answer to fit the one in the book. Later a similar problem appeared on a test. She missed it because she had not asked how it should be done. She had not communicated her confusion to the instructor.*

- *Henry was studying for a mid-term math exam with a group of students. Several members of the group, including him, revealed their confusion about a certain kind of problem. They tried to work on it together and had been doing so unsuccessfully for about 5 minutes. Then another group member interrupted them and said, "I don't see why we have to waste our time with this one when it's obvious and we need to move on to the hard ones." Students who had been working on the first problem became silent. After all, he was probably right. This was an easy one and they should be able to figure it out on their own. The individual who spoke up then steered the group toward the topics of interest to him. The mid-term was difficult and no one in the group, not even the vocal student, did very well. If the quiet members of the group had communicated their needs, the group study experience might have taken a different course.*

Exercise: ASSERTIVENESS RATING

Directions: How effective are your own assertive communication skills? Here is a list of math related situations. Read these and circle all the words which best describe your behavior and feelings.

1. Asking the instructor a question during math class when you are confused.

   Would do — Would feel anxiety

   Would not do — Would feel no anxiety

2. Telling friends, family that you cannot help them, talk to them, etc., because you are studying math

   Would do — Would feel anxiety

   Would not do — Would feel no anxiety

3. Going to see a math instructor during office hours to discuss some math-related concerns you have.

   Would do — Would feel anxiety

   Would not do — Would feel no anxiety

4. Requesting that another person not make "put-down" remarks about your math abilities.

   Would do — Would feel anxiety

   Would not do — Would feel no anxiety

5. Asking a classmate to study with you when you feel you need some help.

   Would do — Would feel anxiety

   Would not do — Would feel no anxiety

6. Asking an instructor to explain his/her grading of certain items on your test.

   Would do — Would feel anxiety

   Would not do — Would feel no anxiety

The more times you said you would be likely to respond in this way, the more skills you have acquired in assertive communication. If you indicated you either would not do something or you would do something but still feel anxious, you could benefit from further skill training.

Most of us are "situationally" assertive. This means that in some situations or with some people we can easily express our feelings and concerns very clearly; in other situations or with other people we find it more difficult. We can each make an ordered list of situations that are difficult for us. That is, we could place the hardest on the top and the easiest on the bottom, with other situations in between. One person might put "asking for help" on the top of his or her list and "complaining about the grading of a problem" at the bottom. For another individual this order might be just the reverse. This means that we will each have different priorities as far as learning new skills. You learn assertive communication skills in the same way you learn other skills, through practice. Here are some sample situations to consider.

Exercise: DEVELOPING ASSERTIVENESS SKILLS

Directions: In each situation described, write down the exact words you probably would use in your response.

1. The math class is finished for the day and you did not understand the concepts being discussed. You go to the instructor and say,

   _____

   _____

   _____ .

2. The math instructor subtracts two points from an answer you gave on a test. He or she gave full credit to the same answer on another student's paper. You show your test to the instructor and say,

   _____

   _____

   _____ .

3. Your roommate is a math whiz. He or she is trying to help you with one kind of problem you can't seem to figure out. After hearing the explanation three times you still don't understand. He or she says, "What is your problem? If you haven't gotten it by now, you'll never get it." You say, _____

   _____

   _____ .

4. A classmate asks you to study with him before the next math test. You feel that this would be unproductive since he is even farther behind than you are. You tell him, _____

   _____

   _____

5. It is the day before a math test. A relative calls and says she would like to bring her children over for you to watch for several hours because she needs to do some vital errand. You tell her. "_____

_____

_____

_____ .

Now test the responses out with a classmate or other person. How does each one sound to you? How do your responses sound to the other person? Here are some questions to ask yourself in evaluating your practice responses.

1. Was I making excuses? (If so you may get caught. One excuse often leads to a long list of excuses. If the other person is very persistent, you will run out of excuses and end up doing what you don't want to do.)

2. Was my response honest, or did it sound like I was lying or hedging on the truth? (The goal is to learn to be honest and direct, not to lie well.)

3. Did I sound as if I were feeling comfortable? (If you were feeling uncomfortable, your response may not have been the best one for you; the other person may sense that and use it to advantage.)

4. What was the rest of my body communicating—confidence, fear, anger? How did my voice sound? Where were my eyes directed? (Your words and your body language should convey a consistent message; otherwise it will not be clear which message to believe.) One practice exercise does not make you assertive in math (or other) situations. Here is a way to apply what you learned from the work above.

Exercise: APPLYING ASSERTIVE SKILLS
Directions: Fill in blanks below as best you can.

1. Write down a school-related situation in which you would like to become more assertive. _____

_____

2. What would you like to do differently in this situation? _____

_____

3. Write down what you want to say, the assertive communication you would like to use in this situation. _____

_____

Now practice this. You may want to try it with a friend. Stand before a mirror and say it. Say it into a tape recorder. Then ask yourself, "How did I look? How did I sound?" If you are not satisfied that this statement is really assertive or that you are really communicating what you want, try something else. Then practice your new assertive communication. Keep modifying until you are satisfied. Some general tips on communicating assertively are:

1. *Be clear* in your own mind about what you want to have happen. For example, if you believe your test was graded unfairly, what do you want? Do you want the instructor to look at it again? If a teaching assistant or student aide graded your test, do you wish the instructor to read it this time? Know in advance *exactly* what you want to result from this situation.

2. *State* what you want to have happen. You cannot assume that your desires will be known, no matter how obvious they seem, unless you communicate them very clearly and precisely. For example, say to the instructor, "I don't understand your grading of this problem. Would you show me why my answer is not correct?*

3. *Be conscious of the nonverbal messages* you are communicating. Correct any obvious problem, e.g., smiling when you are telling your tutor you are unhappy with the way he or she is treating you.

4. *Expect a positive performance* from yourself. That is, expect that you will be able to communicate in a confident, clear manner. (Refer back to the section on self-defeating thoughts if you have forgotten the importance of replacing negative thoughts with positive ones.)

5. *Don't give up* if it is important to you. You may have to try your communication more than one time before you get your point across. The "broken record" technique is sometimes useful. Simply repeat the same statement over and over again, until the other person hears what you are saying.

---

*This may sound weak but we recommend avoiding using language that might put the instructor on the defensive.

STRESS-REDUCTION SKILLS

The topics in the previous two sections are closely connected to stress reduction. Self-defeating behaviors can cause stress even though you may not be aware of it at the time. The inability to communicate what you are feeling or to ask questions when you need information is also a source of stress. When you have developed these skills you have moved a long way toward being more comfortable.

Here are various other techniques or activities you can apply that have the potential to reduce stress in math encounters. These include:

1. timing your math encounters appropriately,

2. relaxation exercises,

3. meditation, and

4. programmed worry.

Timing Your Math Encounters

A widely quoted expert on stress, Dr. Thomas Holmes,* believes that stress and change are closely linked. Holmes has developed a scale of life changes. It includes such events as death of a spouse, taking a vacation, divorce, change in financial state, change in responsibilities at work, celebrating holidays. He has gathered evidence to show that the greater number of life changes a person experiences (he lists 43 possibilities), the greater the chance of serious illness. The implications of this research for a math anxious person are obvious. You should avoid signing up for a math course if you are already involved in a series of stressful events. Time your formal math encounters so they occur in a relatively calm, stable period in your life. This is not to encourage you to postpone enrollment in a math course until *every* aspect of your life is in order and you are completely at peace. This would be another way to avoid math encounters forever. Do delay, however, if some major aspects of your life are out of order. You want your next math class to be the positive experience that it can be if your life is not filled with other stressful happenings at the same time.

Relaxation

People who pursue a regular exercise program, jogging, swimming, bicycling, claim that there is no better way to reduce tension and to provide a sense of well being. Strenuous exercise is a welcome contrast to the

---

*Holmes, T.H. and R.H. Rahe. The social readjustment rating scale. *The Journal of Psychosomatic Research* 11: 213–218, 1967.

Delay your math encounters if some other major
aspects of your life are out of order.

MOVE IT
Y'RSELF

purely mental work of studying. If you exercise before you study it can
help prepare your mind. As a break from study it can help you relax. It
can also place distance between you and a particular problem you have
been worrying about.

Another tool for reducing stress is to practice what are called progres-
sive relaxation techniques. This expression refers to a process of moving
from the lower part of your body to the upper part, relaxing each muscle
in a systematic way. Here is how it is done:

- Sit in a comfortable chair or lie on the floor.

- Close your eyes . . . keep your head centered and your palms up.

- Begin by concentrating on your feet. Tense your feet and hold
  them in a tense position for at least five seconds . . . feel the
  tension fully . . . then relax and notice the contrast. Try this again.

- Continue to tense and relax all parts of your body . . . your calves,
  thighs, abdominal muscles, chest, hands, arms, shoulders, neck,
  and face.

- After you are relaxed, keep perfectly still . . . let yourself sink even deeper into your chair.*

Once again you must be aware that this is a skill that must be repeated frequently in order to have long-term effects. Spending 10 to 15 minutes a day in learning this technique will eventually enable you to relax whenever you choose. You might use it just before or during a test.

Meditation

There is a very good introduction to meditation in the book *How to Meditate* by Lawrence LeShan. It gives an overview of the purposes, techniques, and benefits of this activity. Among the reasons LeShan gives for meditating are to become "more at ease with ourselves, more able to work effectively at our tasks and toward our goals . . ." and to become ". . . less anxious."† He describes many types of meditations. One of the most basic is breath counting. In this meditation, you first find a position in which you can be comfortable for 15 minutes. You place a clock within your range of vision. Then you pay attention to your breathing, counting each time you exhale. Do this in sets of four; start at one again after you have exhaled four times. LeShan points out that what you discover immediately is how much your mind wanders; how difficult it is to focus your attention on just one thing. Rather than letting this lack of concentrative abilities worry you, he advises you just to keep on. Scheduling time for regular, daily meditation will be helpful in bringing about an "increase in mental awareness and alertness and a decrease in physiological tension."‡

Programmed Worry

This is a technique that prevents you from draining your energy through prolonged worry. Let's say that you have been told to expect a math test in one week. You are worried about this test because you did not perform well on the previous one and are not sure you really understand the material. You know that worrying interferes with your being able to study but you don't know how to stop. Many times during each day

---

*Galarosa, Annie, Ann Oxrieder and Marilyn Weckwerth. "How to Take Tests." Seattle: Seattle Central Community College, 1975.

†LeShan, Lawrence *How To Meditate*. New York: Bantam Books, 1974, p. 3.

‡LeShan, Lawrence *How To Meditate*. New York: Bantam Books, 1974, p. 31.

you think about the upcoming test. Not only do you feel anxiety over it but you are now becoming worried about how much time you are spending worrying. You know you have to get down to work but it is very hard to clear your mind of these interfering thoughts. Programming your worry is one approach you might use here. Set aside several 10-minute blocks a day when you will worry, e.g., when you are riding the bus or waiting for dinner to cook. Do this often. Just give yourself up to worry for these regular periods. Think of everything that could possibly go wrong. You could fail the test, perhaps fail the course, feel like a fool, never get the job you want, and on and on. Don't give up. Worry steadily for the entire time you have set. In a short time you will be assigning yourself to less frequent worry periods. Finally with this kind of regular, intense absorption in your concern you will exhaust your ability to worry about it.

CONCLUSION

There are many benefits to be gained from acquiring a whole range of self-management skills. Even though skill building is time consuming the rewards more than justify the efforts. These rewards include (1) allowing you to control some new aspects of your life, (2) increasing your sense of responsiblity for what happens to you, and (3) making you feel better, less stressed, more confident in your abilities. Again, start small. You did not learn to ride a bike, write your native language, or drive a car all at the same time or even in a short time. Skill building takes a long time and requires that you take on one task at a time. These relatively small steps will eventually add up to significant change.

6

PROBLEM SOLVING — PART I

This chapter will

- *define problem solving,*
- *describe blocks to that process, and*
- *help you identify your strengths in problem solving.*

DEFINITION OF PROBLEM SOLVING

Problem solving deals with three processes. These are:

1. coming up with a new answer to an old problem,

2. getting an old answer (but new to us) to an old problem, or

3. recognizing new problems and also possible new solutions.

At one time everyone believed that creative problem solving was an ability you were either born with or born without. This is not a common belief today.

One psychologist* speaks for many people when he says that there is an inherited trait called "creativity." This has characterized many geniuses throughout history and cannot be taught. But there are also certain problem solving skills and personality characteristics that can be learned or developed. Another psychologist believes that creativity is "the universal heritage of every human being that is born. . . ."† Many mental health experts‡ think creative problem solving skills go hand in hand with being healthy individuals who develop their potential. Under the right conditions, they argue, each of us can become a creative problem solver. To these experts, the right conditions are those in which a person is given support in a criticism-free environment.

BLOCKS TO PROBLEM SOLVING

Psychologist Paul Torrance# describes five conditions in our society that keep us from thinking creatively. *First* is our stressing the importance of success. Often creative solutions require taking risks; yet it is hard to

---

*Attributed to Ausubell by Bruce P. Holleran and Paula R. Holleran, Creativity revisited: A new role for group dynamics. *The Journal of Creative Behavior* 10 (2): p. 130, 1976.

†Abraham Maslow, "Creativity in Self-Actualizing People." In Harold H. Anderson (ed.) *Creativity and its Cultivation.* New York: Harper and Brothers, Publishers, 1959, p. 84.

‡See articles by Erich Fromm, Abraham Maslow, Rollo May, and Carl Rogers in Harold H. Anderson (ed.) *Creativity and Its Cultivation.* New York: Harper and Brothers, 1959, pp. 44–95.

#Torrance, E. Paul. Nurture of creative talents. In Ross L. Mooney and Taher A. Razik (eds.) *Exploration in Creativity.* New York: Harper and Row, 1967, pp. 185–195.

take risks when the consequences may be failure. Unfortunately there are few social rewards for failing and many of us become afraid to try. Consider this account of how Thomas Edison solved the problem of finding a chemical that dissolves hard rubber. Other scientists had used extensive theory and formula, but Edison began by dipping a small piece of hard rubber in each of the chemicals available in his storeroom. Each time the experiment failed, he claimed he had narrowed down the problem. Finally patience won out and he discovered an agent that dissolved the rubber. Sometimes genius is spelled P-E-R-S-I-S-T-E-N-C-E.

A *second* social condition is the strength of peer influence on people's behavior. Starting in the fourth and fifth grades, children tend to become less creative. This may stem in part from the need to be like everyone else. A *third* influence is the pressure to accept events without questioning. Often asking questions is not encouraged, particularly in the classroom. This too prevents creativity. *Fourth*, placing students in certain classes on the basis of sex may hamper their abilities to do creative problem solving in a nontraditional area. For example, girls are directed into the arts and boys into science. A *fifth* problem is our need to get things done quickly to meet deadlines. Each of these norms may restrict our abilities to become creative problem solvers.

Besides these social conditions there is a second influence on our abilities to solve problems, our own attitudes. One expert* listed attitudes that get in the way of creative problem solving. These include:

1. using only critical, negative thinking;

2. relying only on old solutions and habits to solve new problems;

3. feeling fearful of looking stupid; and

4. refusing to become really involved with the problem.

One cannot be afraid, negative or uninvolved. Creative problem solving requires a strong commitment.

Two other feelings that restrain our problem-solving abilities are anger and distrust. Sometimes the anger is toward the math problem, "This is the stupidest problem I've ever seen." Sometimes it is toward the individual who asked us to do the problem, "That so-and so! He's ruining my weekend with these silly math problems!" The anger may also be directed toward ourselves, "How can I be so dumb," "Why does it take me so long?" In any of these cases the anger may keep you from solving the problem. Since feelings of anger occur spontaneously, it is

---

*Osborn, Alex. *Applied Imagination.* New York: Charles Scribner's Sons, 1957.

usually a waste of energy to try to ignore them. The best remedy is to temporarily remove yourself from the source by taking a short break.

When we don't see an immediate solution to a problem, we sometimes begin to feel distrustful, to suspect that the problem contains a mistake or a misprint, and that it cannot be done. You may have even taken math classes where many of the problems involved silly tricks. Distrust easily turns to frustration, "I know there's a trick here and I just can't see it." This is similar to our reaction to being unable to find quickly the right piece in a jigsaw puzzle. We assume it must be missing. The frustration leads back to anger. The majority of math problems we encounter are not tricks or misprints. Our problem solving effectiveness will depend on our ability to accept this fact.

Our energy is best spent on trying to solve the problem, rather than looking for the trick or error. We need to recognize our feelings of anger and distrust and accept the fact that they will probably continue to occur. We can then make an effort to move beyond them as quickly as possible.

The problem solving process has been described in different ways. The most basic way identifies four stages of the solution process.* These stages are called *preparation, incubation, illumination* and *verification.*

Preparation is the phase where initial information is gathered. In this stage you become very familiar with the problem. You study it carefully and make sure you understand it. You may not have enough information to solve the problem yet. Reading and talking to others permits you to expand your base of knowledge, which will be critical to understanding the problem thoroughly. After you have acquired sufficient information you "play around" with the problem for awhile. At this point you may develop some tentative solutions.

In the *incubation* stage you turn away from the problem to engage in other activities. These other activities may be getting a good night's rest, getting some exercise, taking a shower, or anything that takes you away from concentrating on the problem. During this period, some of your mental blocks and inhibitions are released or lessened.

During the *illumination* stage you may suddenly experience insights into the solution of the problem. You often see things in a different way; an approach you had never considered becomes apparent to you. When this happens people frequently shout: "I've got it!" "That's it!" "Aha!" or some similar exclamation.

---

*McPherson, J.H. "The People, The Problems and the Problem-Solving Methods." In Sidney J. Parnes, Ruth B. Noller and Angelo M. Biondi, *Guide to Creative Action.* New York: Charles Scribner's Sons, 1977, p. 147.

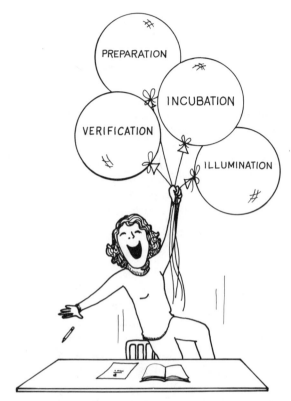

Solving your problems well can give you a lift.

The work of refining and testing your solution comes in the *verification* stage.

We need to remember that each of us has already developed many problem-solving skills. We could not have reached adulthood without them. We figure out why the car won't start and how to get to our destination without it. We can design an evening's entertainment that costs little money, or we manage to have the meat, vegetable and salad all ready for dinner at the same time. When dealing with math problems, the difficulty is that our fear and lack of confidence make us forget our skills entirely. We act as if we have no relevant traits or experiences upon which to call in solving the problem.

The following exercise may serve to remind you of the problem solving skills you already possess.

Exercise: PROBLEM SOLVER INVENTORY

Directions: Place a check mark by those personality traits that you believe describe you.

_____ Flexible

_____ Open-minded

_____ Curious

_____ Like to learn new things

_____ Like challenges

_____ Enjoy trying new things

_____ Don't make quick judgments

_____ Like to figure out things by myself rather than having some-one else tell me.

Now do the same thing with this list of skills.

_____ Read carefully

_____ Look up new words in a dictionary

_____ Compare two or more things; notice similarities and differences.

_____ Notice details

_____ Separate important from unimportant information

_____ Plan ahead

_____ Break complex problems into simple parts

_____ Organize information

Finally, here is a list of experiences. Check any of these that are similar to ones you have had.

_____ Read, interpreted, and carried out directions accompanying a sewing pattern.

_____ Designed, planted, and tended a vegetable garden

_____ Built a model

_____ Built or made something with a plan of your own design

_____ Assembled a piece of equipment

_____ Repaired a piece of equipment or household item

_____ Solved a photography problem related to lighting, depth of field, or shutter speed

_____ Solved jigsaw puzzles

_____ Played board games

_____ Adapted a receipe (to ingredients on hand, or a different
number of servings)

Look back over these checklists. Very likely you have checked several
items in each category. Since the lists contain personality traits, skills,
and experiences that are useful in problem solving, then you are not
starting totally without skills. You already have developed some problem-
solving abilities and now you are going to build on these to learn more.

Exercise: PROBLEM-SOLVING ABILITIES SUMMARY
Directions: Use the lines below to write a brief summary of your
strengths as a problem solver. If possible describe a real situation in
which you solved an intellectual, i.e., a nonemotional problem.

_____

_____

_____

_____

_____

_____

With this background on the definition and processes of problem
solving and your own strengths in this area, you are ready to move on to
specific problem solving techniques discussed in the next chapter.

FURTHER READING

Campbell, David. *Take the Road to Creativity and Get Off Your Dead End*. Illinois:
Argus Communications, 1977.

some useful techniques

7

PROBLEM SOLVING — PART II

This chapter discusses various strategies for problem solving including
- *beginning with a positive attitude,*
- *reading carefully,*
- *keeping an open mind,*
- *organizing the problem,*
- *looking for patterns,*
- *using analogy,*
- *working with simpler cases,*
- *using elimination,*
- *visualizing the problem,*
- *playing by the rules,*
- *determining approximate answers,*
- *choosing a starting point,*
- *checking your intuitions, and*
- *taking time to get results—incubation.*

These techniques are useful in finding solutions to mathematical problems. Not every method will be helpful on any particular problem, but part of being a good problem solver is deciding which strategies might be useful. Throughout the chapter, sample problems are introduced to illustrate the topic under discussion. Do not feel you are expected to solve every problem by yourself. We are interested in strategies, not just solutions. Although some of the answers are given as a way to understand the technique, in many cases the solution is left incomplete. Feel free to work on the problem further, perhaps posing it to friends and associates as well. The solutions are given in Appendix A.

BEGINNING WITH THE RIGHT ATTITUDE

Approaching a problem with positive feelings is essential. You must have *confidence* in your own ability. You must give the problem total *concentration* for a period of time, and you must be *committed* to solving it. These seem to be strong words about something as trivial as working on a math problem that may hold little real value for you. But it has been shown that confidence, concentration, and commitment are crucial to good performance. This is true not only in math, but in most other fields as well. It has been demonstrated that athletic performance can be dramatically increased by *thinking* right. That is why many college and professional teams employ a psychologist as well as a coach. Reserves of strength

and ability that aren't even known to exist can be called on when there is intensity and commitment.

To the extent that you learn to manage and control your attitude and feelings about problem solving, you will increase your rate of success. If you begin a problem "knowing" it is too hard for you, then you will probably fail. This is a self-fulfilling prophecy. If you do not force yourself to concentrate, read carefully, and attent to detail, the problem will become more complicated. And finally, if you don't really care whether you solve it or not, then you will probably quit before a solution comes to you.

READING CAREFULLY

Questions must be read carefully and slowly, giving attention to each word. One of the chief causes of inaccurate problem solving is trying to consume the information too quickly and in doing so missing key points.

example 1: A kennel contains forty dogs. Among them are 12 German Shorthairs, 10 Great Danes, 5 Cocker Spaniels, 7 Collies, and 4 Terriers. How many dogs are in the kennel?

If your answer was 38, then you correctly added up the total for the breeds mentioned. An accurate reading of the first sentence however informs us that there are 40 dogs, not 38. The second sentence does not contradict that since it only says "Among them are. . . ." This example was deliberately designed to mislead you. But similar mistakes occur commonly in reading any math problem. You will tend to miss crucial information if you read questions too quickly or too passively. A useful technique is to first skim the problem quickly to get the general content. Then read slowly, stopping after each sentence or phrase and asking yourself if you really understand its meaning. Can you rephrase it? As you go along try to jot down relevant facts for later reference. Reread the problem at least once to see if you missed anything. You will be amazed at how much additional information can often be found.

Since any word or symbol may be important, it is essential to understand the precise meaning of each. Some words may require going to a dictionary or reference book. Do not skip over things you're not sure of.

Some problems are poorly defined or are unclear. If this is true of the problem you are working on, you may have to seek additional information and direction. Or you may just have to work the problem in several different ways, one way for each interpretation. In other cases the problem may contain irrelevant information, as did Example 1. Part of your job is to discard useless and distracting material.

KEEPING AN OPEN MIND

Another obstacle to problem solving comes from the unique history and experience each of us brings with us to the task. Sometimes the richness of our past is helpful in understanding the problem, but unfortunately it can also get in the way. Occasionally we miss solutions because our subconscious has already eliminated them from our consideration.

example 2:   Do they have a 4th of July in England?
   The answer is clearly yes, just as they have a 3rd and a 5th of July. If we have trouble answering this question, it is because we associate the 4th of July with the American holiday of Independence Day.

example 3:   A father and his son were involved in a terrible automobile accident. The father was killed and the son critically injured. As the son was being rushed into the operating room at the hospital, the surgeon said, "I can't operate on this boy. He is my son." Explain.
   The explanation is quite difficult if we assume that surgeons are always male. As more and more women enter new fields, the solution will become very obvious.

example 4:   A lake is in the form of a square, with a tree planted at each corner. How could you enlarge the lake, so that the area would double, the lake would remain a square, and you would not disturb any of the four trees?

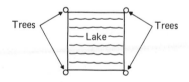

The solution here involves changing the orientation of the lake. Many people fail in their attempts to solve the problem because they can only see the square in the position given. This is a restriction that is not imposed by the problem, but rather by the problem solver. The problem solver has closed his or her mind and has eliminated other possibilities. Can you see how to enlarge the lake before turning to Appendix A for a solution?
   Unfortunately there is nothing foolproof that can be done to avoid making unwarranted assumptions and to keep an open mind. Sometimes

just knowing that this difficulty exists is helpful. You can also try writing down your conscious assumptions and then checking to see if any are unnecessary.

ORGANIZING THE PROBLEM

Some problems can be solved merely by sorting through the clutter. The information as it is presented may be totally unworkable, and can only be grasped as it is rewritten and reordered. The structure you provide may be very formal or may be as simple as keeping track of details on a separate piece of paper. Restructuring the information in the problem is an important tool to use, particularly when you feel stuck. Sitting and staring at the same clues, in the same form, seldom leads to inspiration. Almost anything different that you can do with the material will be helpful.

example 5: Abbott, Brown, and Casper are a detective, an entomologist, and a farmer, although not necessarily in that order. Abbott was the proud father of healthy twins yesterday. Casper has a deathly fear of insects and won't even get close to one. The farmer is getting worried because he and his wife are getting old and with no children to help he won't be able to run the farm for many more years. Casper, the bachelor, especially likes brunettes. What is the occupation of each of the three men?

There is only one paring that satisfies *all* the facts given. Have you found it? Do you understand all the words used in this example? If you don't know what an entomologist is, you can't solve the problem. Find this definition first before you try anything else. How are you organizing the information that is given? Try jotting down key facts as you reread the problem again. Each person will probably do this in a slightly different way, but some system is essential. An illustration of what might be noted is shown below.

Abbott: father (not farmer)   farmer: married, no children
Brown: ?   detective: ?
Casper: fears insects (not entomologist)   entomologist studies insects
      bachelor (not farmer)

From this information we can conclude that Casper is neither the entomologist nor the farmer. The only choice left is the detective. Then, since Abbott is not the farmer, and we now know not the detective, he must be the entomologist. Brown is the farmer. Notice how much useless information was given in the statement of the problem. Sometimes it is difficult to know which details to ignore.

The more complicated the problem, the more organized you need to be in dealing with the information. Outlining it, making lists, and rearranging information may all be helpful. The important thing is to begin to manipulate the given facts. Problems numbered 37, 38, and 39 in the next chapter, illustrate the necessity for these techniques.

LOOKING FOR PATTERNS

Looking for patterns is really the essence of what scientists call "inductive reasoning." This is the method used in serious research in both the physical and social sciences. Doctors use it as they seek to isolate the causes of a particular disease. Psychologists use it in trying to explain a certain kind of behavior. The key elements are:

- collecting data,

- organizing it systematically,

- discovering relevant relationships or patterns, and

- checking "discoveries" with more data.

Often solutions to math problems can be found by noticing patterns or relationships. How successful you are at this depends on how skillfully you organize the information and to some extent on how familiar you are with the subject in question. It may also depend on the length of time you are willing to spend making observations.

example 6: Number patterns appear frequently and have some interesting uses. Try to decide what common pattern exists between the given numbers in each example and then fill in the next two entries.

a) 1, 4, 7, 10, 13, ———— , ———— .

b) 1, 2, 4, 8, 16,   ———— , ———— .

c) 1, 1, 2, 3, 5, 8, ———— , ———— .

d) 1, 4, 9, 16, 25, ———— , ———— .

Part (a) is an example of an arithmetic sequence. This is a sequence where a fixed number is added to each term to get to the next one. Part (b) is a special geometric sequence called the binomial sequence. (Can you decide why it is called binomial?) It is the basis of internal operations of computers. Part (c) is the famous Fibonacci sequence whose patterns occur in such diverse places as the family tree of a male honey bee, the spirals of nature, and the proportions of the "golden ratio,"* part (d) is called the sequence of squares. The correct completions are given in Appendix A.

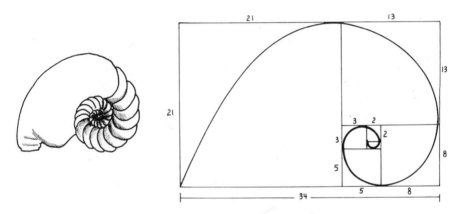

A cross section of the snail-like nautilus
displays the numbers of the Fibonacci sequence.

example 7: There is a pattern in the lines of numbers shown on the next page. Fill in the next line in the pattern. If you don't see the whole pattern, write in whatever pieces you do see.

---

*The golden ratio is utilized in art and architecture as the proportion that is most pleasing or beautiful.

```
                    1
                 1     1
              1     2     1
           1     3     3     1
        1     4     6     4     1
     1     5    10    10     5     1
  1     6    15    20    15     6     1
```

This number arrangement is known as Pascal's Triangle, named after Blaise Pascal, a French mathematician of the 1600s. Although Pascal showed its application to the theory of probability, people had been fascinated by its properties long before his time. The triangle appeared in the Orient at least a thousand years before. If you look carefully you can discover both the Fibonacci and binomial sequences (introduced in Example 6) inside the triangle.

USING ANALOGY

Insight into a problem can come from studying another one that is similar in some way. In mathematics, for example, points, lines, and planes are considered figures that are analogous or similar. This is based on the fact that a point is the simplest one dimensional object; a line is the simplest figure in two dimensions; and a plane is the simplest figure in three dimensions. If you were given a mathematical problem that involved planes, it might be helpful to think of a similar problem dealing with lines, or even points.

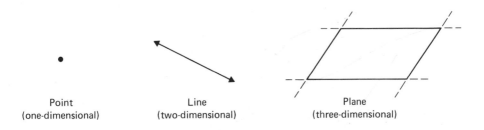

| Point | Line | Plane |
| (one-dimensional) | (two-dimensional) | (three-dimensional) |

The ability to see analogous situations is very helpful in mathematics. It may help you recognize that a current problem is merely a different version of one you already know how to solve. This ability can be enhanced and developed with practice. Many standardized tests use analogy to measure certain kinds of aptitudes of intelligence. Practicing for these tests will usually improve your score, making you actually score "smarter." In general, drill does not improve thinking, but it does give experience which is beneficial in taking certain types of tests. For this (and other reasons) such tests are sometimes considered useless or invalid as measures of basic intelligence.

example 8: Here are some typical analogy questions found on standard IQ (intelligence quotient) tests. Try to select the one word or symbol that best completes the analogy. In (a), (b), (e), and (f), it may be helpful to describe to yourself in words, how the first two parts of the analogy are similar, or what relationship exists between them. In (c) and (d) try to list the things common to the figures.

a) Which of the five items makes the best comparison? Milk is to glass as letter is to:

      stamp   pen   envelope   book   mail

b) Which of the five designs makes the best comparison?

c) Which of the five designs is least like the other four?

d) Which of the five designs is least like the other four?

e) Which of the five designs makes the best comparison?

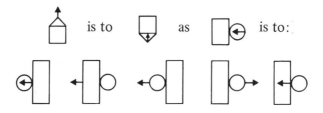

f) Which of the five designs makes the best comparison?

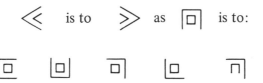

WORKING WITH SIMPLER CASES

A problem may be difficult because of the quantity of information in it.
The method needed to solve it may be hidden by irrelevant details.
Sometimes a simpler version of the same problem has a solution that
you can readily see. The method used to solve the simpler problem can
then be used to operate on the more complicated one.

example 9:   A car travels 36 miles in 45 minutes. Traveling at the same
speed how much distance can it cover in $4\frac{1}{2}$ hours?
   This problem is complex for two reasons: the numbers are a bit messy
and the times are measured in two different units, minutes and hours.
Try to think of a similar problem that you *could* solve. For example:

> A car travels 10 miles in 1 hour. How far would it travel
> in $4\frac{1}{2}$ hours?

Is this easier to solve? The car would go four and one-half times as
far in $4\frac{1}{2}$ hours as it would in 1 hour. If it goes 10 miles in one hour it
should go one half that distance, or 5 miles in a half hour. So the distance
traveled would be $(4 \times 10) + 5 = 45$ miles.

Now go back to the original problem. If we can figure out how far the car goes in 1 hour then we can tell how far it goes in $4\frac{1}{2}$ hours. We might argue that 45 minutes is 3/4 of an hour. So in 1/4 hour the car would go only a third of this distance. This would be $36 \div 3 = 12$ miles. Then in an hour it would go $4 \times 12 = 48$ miles. In one half hour it would go $48 \div 2 = 24$ miles. Finally in $4\frac{1}{2}$ hours it would go $(4 \times 48) + 24 = 216$ miles. Do you see how the last part of the solution is identical to that used in the similar, simpler problem? There are of course many other methods that might have been used to find the solution. If others seem easier or more reasonable to you don't hesitate to use them.

Sometimes a combination of techniques is useful. In the following example, considering simpler cases *and* observing the patterns that develop are both helpful.

example 10:   Imagine folding a piece of paper in half, and then in half again, and repeating this process until the paper has been folded a total of 10 times. How many thicknesses of paper would you have?

The complexity here arises from trying to picture a total of 10 folds. First try picturing how many thicknesses there would be if you were to fold the paper in half just once? How many thicknesses if the paper were folded twice? Three times? Write these results down and see if there are any patterns that emerge. If so, make a guess at how many thicknesses there would be after 10 folds. The correct result is $2 \times 2 \times 2 \times 2 \times 2 \times 2 \times 2 \times 2 \times 2 \times 2$, which is 2 multiplied by itself 10 times. A mathematical shorthand for this expression is 2^{10}. The multiplication can be done either longhand or with a calculator, yielding a value of 1024 thicknesses.

USING ELIMINATION

Often it is easier to determine what a solution *is not*, than to decide what it *is*. This has already been illustrated in Example 5. We solved this problem by eliminating two of the three possibilities for Casper, leaving only one choice. Problems that at first appear impossible because there seems to be insufficient information or too many methods of attack, often lend themselves to solution by this method. One of the key steps in elimination is to recognize *all* of the possibilities. Sometimes it is help-ful to list them all first and then cross each off as you dispose of it as a possibility.

example 11: You have eight golf balls. Seven of them are identical in weight but one of them is slightly heavier. You are given the use of a balance scale to determine which is the odd (heavier) ball. Describe a procedure you would use to find the heavy ball.

If you're not sure how a balance scale works, you'll need to learn about that first. After you have devised a procedure to find the heavy ball, determine how many times you used the scale. Your answer is probably three or four. One of the more popular solutions to this prob-lem is as follows.

First place four balls on each side of the scale, weigh them, and save out the heavier four. Split these up, putting two on each side of the scale. Take the heavier group of two balls and split them up putting one on each side of the scale. The side that goes down on this weighing contains the heavy ball. This took *three* weighings. A picture of the weighings with the balls numbered for identification is shown below.

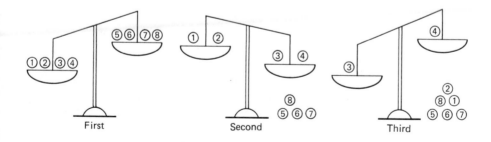

First   Second   Third

In this sample situation, ball number 3 turned out to be the heavy ball.

example 12: Consider the situation in Example 11 again. This time you are being asked to find the heavy ball, but must use the balance scale *only twice*.

The difficulty with our previous solution is that it required three uses of the balance scale, not two. Do you see a method that would require only two weighings? Many people feel stuck at this point. When their first try fails they find it hard to consider other alternatives. In the solution to Example 11 we began by putting 4 balls on each side of the scale. How else might we begin? Let's list all the possible combinations of balls that would be reasonable on the *first* weighing. They are:

- place 1 ball on each side, reserve 6 balls unweighed;

- place 2 balls on each side, reserve 4 balls unweighed;

- place 3 balls on each side, reserve 2 balls unweighed; and

- place 4 balls on each side, reserve no balls.

We have already tried the last possibility in Example 11, placing four balls on each side for the first weighing. We saw that a total of three weighings was necessary to determine the odd ball. It remains to try the other three possibilities. It helps to keep track of your results at each weighing and to consider all possible outcomes.

Many times we feel stuck long before we've really tried all our alternatives. So one strategy is to list carefully every possibility that occurs to you, however remote. Then explore the alternatives carefully, one at a time. Often what at first seems improbable, will lead to a solution.

VISUALIZING THE PROBLEM

There are various modes in which a problem may be presented. It can be written in sentences as were all the previous examples. It can be read out loud. It can be in the form of a diagram with little additional explanation. Everyone has their own favorite modes through which they can best interpret information—by talking, reading, hearing, or touching. We have only to watch small children enjoying something pretty to decide that their primary mode is touch. When solving problems it is often beneficial to consider the problem in as many different modes as possible. This may involve reading the problem out loud to yourself, drawing diagrams or sketches, or finding physical models and moving them around. Any of these things can serve to make the problem less abstract and more comprehendible.

example 13: You have 20 feet of fencing with which to build a rectangular pen. Find the dimensions of the pen that will have the *largest* possible area.

Note: A rectangle is defined as any four-sided figure whose adjacent sides are at right angles (perpendicular). The area of a rectangle is found by multiplying the length by the width.

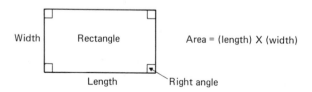

How might we best begin this problem? When geometric figures are involved in a problem, drawing sketches is often useful. We might try sketching rectangles of various proportions and computing their areas. We must be sure that the distance around the outside totals 20 feet.

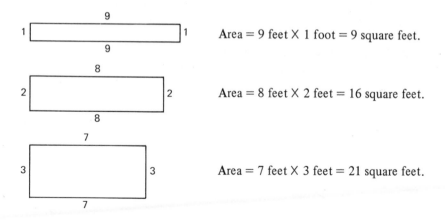

Try drawing other possibilities in this same fashion and computing the areas. You should be able to find a rectangle with an area of 25 square feet. It turns out to be a rectangle that measures 5 feet on each side. An area of 25 square feet is the largest area that can be obtained. It requires the use of more advanced math to *prove* that no larger rectangle exists. At this stage we will *assume* it is the largest possible, because no larger one has been found. This is not a desirable situation in mathematics but is satisfactory for our current discussion.

Although sketches are usually essential in a geometric problem, many people don't bother to draw them. Rather, they try to picture mentally what is described. This is not as helpful as putting it down on paper. It is advisable to draw as many pictures, from as many different perspectives, as you can. Even problems that are not geometrical often become clearer when they are translated into pictures.

example 14:   Two cars, call them A and B, begin driving at the same time. Car A begins 20 miles ahead of Car B. Car A is traveling at 50 mph (miles per hour) and Car B is traveling at 60 mph. How long will it take Car B to overtake Car A?

This problem can be done without pictures, but drawing them never hurts, and might help. We begin with a blank line and indicate where each car starts and the direction of travel. One version might look like the sketch below. This picture is much like a ruler measuring miles instead of inches. It shows Car A beginning 20 miles ahead of Car B and indicates which way the cars are traveling. We have shown where each car would be at the end of one hour. Car A will have moved 50 miles to the right of its starting point and Car B will have moved 60 miles to the right of its starting point. Now you put a large X below the line where Car A will be at the end of the second hour. Do the same above the line for Car B. Can you now answer the question of how long it takes Car B to overtake Car A? Can you also tell what the total distance traveled by each car would be at the moment they meet? More complicated problems of this sort often require using methods of algebra to get precise answers. But diagrams such as this are still helpful in visualizing what is happening.

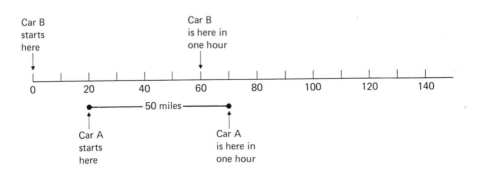

example 15:   Using three matchsticks, form a triangle with equal length sides. Now using only six matchsticks, form 4 triangles. The sides of the triangles must all be the same length and you must not let any of the matchsticks cross.

While you are working on this problem, try to think of different strategies for drawing or constructing the situation. Have you limited yourself to sketching the picture with pencil and paper. Could this be an artificial restraint? Although pictures may be very helpful, they can also be drawn in ways which are misleading. They can assume things which need not be assumed and cause us to skip over other possibilities. For example doing a drawing with pencil and paper assumes the solution can be found on a flat surface. How about actually using matchsticks, tooth picks, etc. to construct the triangles? Often the very act of moving the objects about can provide insight into solving the problem.

You might want to go back to other problems, like Examples 11 and 12, and ask yourself if there are several strategies that might have been helpful. Is writing alternatives down on paper the only technique that can be used? For many people a more life-like model, such as actual balls and a scale, is more useful.

PLAYING BY THE RULES

When problems get frustrating, it is tempting to add information that was not given, or to change an irritating restriction slightly. In some situations this would be called a creative solution. But as you engage in mathematical problem solving, you must be very careful that your creative solutions are not merely attempts to go around the restrictions given in the problem. The rules may seem contrived, but they are part of the design of the situation. To change them may mean you will miss the point of the problem completely.

For example, in the golfball problem, Example 12, some creative person usually makes the following suggestion to avoid the restriction of two weighings. Begin by placing one ball on each side; then keep adding one to each side until the odd ball is encountered. Notice that is really bending the restraints of the problem. Each time the scale is read should count as one weighing. The process just described could take as many as four weighings unless we just got lucky and found the odd ball right away. Another ingenious method is to use the scale twice (as in the solution given for Example 11) and then take the remaining two balls on the heavy side and hold one in each hand. The odd ball is found by deciding which "feels" heavier. But this is really a third weighing; we are using our hands as the balance scale. Both of these suggestions, though clever and in some situations admirable, are merely devious means to avoid the restrictions of the problem.

DETERMINING APPROXIMATE ANSWERS

It is usually helpful to check that your solutions to problems are reasonable. It was mentioned previously that answers are often submitted that make no sense, regardless of whether the individual knew much mathematics or not. If we heard that 21½ people attended the concert, we would have an intuitive feeling that something was wrong. Yet when we start doing math problems, we often forget that our common sense is still a very important tool. When working numerical problems, it is often helpful to guess at a reasonable answer. Sometimes this approximation actually helps to solve the problem precisely.

example 16: Suppose you put $100 in a bank that gives 6% annual interest. About how much money would you have at the end of two years? Instead of finding the actual value, try to think of what would be reasonable. You would have more than ___?___ but less than ___?___.
  Certainly if you deposit the money in a bank you would expect to get back more than you put in. So the result should be more than $100. Without knowing anything about interest rates and how to compute them (this will be discussed briefly in Chapter 10), you can still probably decide on a reasonable upper limit. It would be very optimistic to expect to double your money in two years, so we might say $200 would be a (high) upper limit.

example 17: An item is on sale for 15% off. Its original price was $39.95. Approximate the reduced price. The item would cost less than ___?___ but more than ___?___.

CHOOSING A STARTING POINT

It is tempting to try to solve a problem by starting at the first fact stated and proceeding in order, step by step to the end. But many times this is impossible. The key to the solution may appear in the last sentence. Therefore it is important to consider the entire problem before plunging in to find a solution. The appropriate place to begin may not be intuitively obvious at first.

example 18: An article requires a *minimum* of four standard coins (U.S. currency) to pay for it. Two of these articles would require a *minimum* of six coins. Three of them can be purchased with two coins, however. What is the price of the article?

The solution to this problem requires you to think of combinations of coins. You might begin with the combinations that would purchase one of the item. After all, that is what the question asks for. However, that means you would have to consider combinations of four coins. Since there are fewer combinations of two things, this might be an easier starting point. Incidentally, if it appears that this problem may have more than one solution, you should go back and read the statements again carefully. Pay particular attention to the word "minimum."

CHECKING YOUR INTUITION

We have all been told at one time or another that guessing is improper in mathematics. However as we have indicated before, it is our opinion that guessing and using intuition is perfectly acceptable. It is, in fact, desirable within limits. Sometimes these skills are quite essential to successful problem solving. For example, intuitive guesses about the size or kind of numbers which are reasonable answers is a good checking procedure. Those who refuse to guess at all, may actually be saying that they won't risk being wrong. Risk taking is essential in problem solving. It would be difficult to even get started on a hard problem if you were afraid of making errors. Solutions to challenging problems normally involve a substantial amount of trial and error. Often something significant is learned through making mistakes.

There is a difficulty with intuitive answers and guesses however. They may overlook critical mathematical properties. There may even be several intuitive answers all of which seem reasonable. The way out of this dilemma is to check all results against *every* detail of the problem. There is nothing wrong with guessing; but there is something wrong with accepting guesses as the final answer.

Many inaccurate solutions given to problems are the result of an incomplete process. You must not allow yourself to simply "see" the solution to the problem part way through, write it down, and go merrily on your way. Answers obtained in this manner are rarely correct.

INCUBATION

The process of incubation has been discussed previously along with obstacles to problem solving. It will be considered again here from a more positive perspective. Math anxiety sufferers often believe they should be able to solve problems immediately. They become discouraged when they can't, and begin to feel their efforts are being wasted. It should be comforting to know that even excellent problem solvers do not usually get results to difficult questions on their first try. When creative people work on a problem and can't immediately find the answer, they keep trying. They continue to work on the task using trial-and-error until fatigue or frustration cause them to put it aside. Then for a period of time the problem is not consciously considered. During this time, which may last from a few hours to several days, the subconscious continues to work on the problem. It is able to restructure the information the conscious mind has fed it. Incorrect sets and directions fade away, making it possible for a new approach to emerge. Sometimes the solution to a problem will jump out while doing totally unrelated tasks. Other times inspiration comes the next time the problem is confronted.

Math teachers should probably take a large share of the blame for our feelings that solutions should be immediate. They present their solutions to students in finished and polished form. They rarely show us their own difficulties with solving a problem for the first time.

Of course incubation isn't helpful until a great deal of work has already been done on the problem. The question must be clearly understood, and all of the relevant facts must be stored away for the subconscious to consider. However, if you've seriously worked on a problem and the solution still doesn't come, perhaps it's time to take a break. This is particularly true if you're becoming frustrated, angry, or tired.

Incubation isn't helpful unless a great deal of work has already been done on the problem.

Put the question aside and return to it later. Time in between can be thought of as bonus study time. The important point is to return to the problem later. When you are fresher you'll be better able to concentrate and you will also have the benefit of the musing that has gone on in the meantime.

FURTHER READING

Adams, James L. *Conceptual Blockbusting, A Guide to Better Ideas*. New York: W.W. Norton & Company, 1974.

Green, J.E. *100 Great Scientists*. New York: Washington Square Press, 1964.

Koestler,

practice problems: now it's your turn

8

PROBLEM SOLVING — PART III

This chapter presents a series of problem solving exercises grouped into three categories

- *reading,*
- *spatial, and*
- *reasoning*

INTRODUCTION

Now you are ready to try your own problem-solving skills on a short collection of questions. You are not expected to solve all or even the majority of the problems given. A few of the questions have answers that you may see almost immediately, but many of them require sustained and serious effort. Part of learning to solve problems is learning that the satisfying feelings that come when you finally see the solution must often be postponed for a time. No attempt has been made to separate the "hard" problems from the "easy" ones or to arrange them in increasing order of difficulty. This omission has been deliberate although we realize that many people are more comfortable with some other system. If problems are supposedly grouped by difficulty, then it is hard to be objective and open as we try to solve them. We *expect* that we should be able to solve the early ones "easily" and may avoid the later ones entirely because we already know they will be impossible. In fact, a problem that is easy for one person may be very difficult for someone else. We suggest you simply choose problems that seem interesting to you. Don't worry about whether they are supposed to be difficult or easy.

There has been an attempt, however, to group the problems into three categories: *reading, spatial,* and *reasoning.* The category describes the primary skill that is used in solving the problems. For example, the reading problems require careful reading with attention to detail. The spatial problems ask you to visualize objects in either two or three dimensions. The reasoning problems, by far the largest section, require careful manipulation or assembly of information to extract what is not at first obvious. Although a few of the latter use some basic arithmetic, most require nothing more than common sense and some mental effort. Remember that even where arithmetic is involved, the key to the solution lies in deciding what needs to be done or what operation has to be performed on the numbers. You can always use a calculator to do the actual computations. (Calculator use is further discussed in Chapter 10.)

Now we are ready to begin. Working on several problems at a time should probably be avoided, because you won't be able to put enough energy into any one of them. Try to pick one problem that interests you and stay with it for a while. Don't concentrate so hard on getting an answer that you forget to notice which techniques or approaches you are using. If you get stuck, go back to Chapter 7 and check to see if there is some other technique you may be forgetting to use. The answers are given in Appendix B. It is suggested that you look there only after you have come up with a proposed solution on your own. Once you've read the answer the puzzle is spoiled. The answer given is only a check to see if you've come out okay. Knowing the answer does not substitute for solving the problem. Besides, you may come up with a better solution, one we have not thought of yet.

Reading Problems

1. A farmer has 12 cows and all but 9 of them die. How many does he have left?

2. Some months have 30 days and some have 31. How many have 28 days?

3. If you had only one match and entered a room where there was a kerosene lamp, an oil heater, and some kindling wood, what do you light first?

4. A rooster is sitting on the peak of a roof and lays an egg. Which way does the egg roll?

5. Is it legal in Washington State for a man to marry his widow's sister?

6. A plane crashed on the border between the United States and Canada. Where were the survivors buried.

7. Three women registered at a motel and paid $30 for the room, splitting the cost equally. Later on the clerk discovered that he had overcharged them. The room should only have cost $25. So he gave $5 to the bellhop to return to the women. The bell-hop (who suffered from math anxiety) couldn't figure out how to split the $5 three ways, so he did the following. He returned $1 to each of the women and kept the remaining $2 for himself. Now each woman originally paid $10 and each then received $1 back. So each has actually paid $9. $9 times 3 is $27. The bellhop has $2. Altogether this totals $29. But originally there was $30. What happened to the other dollar?

8. How many 3¢ stamps are there in a dozen?

Spatial Problems

9. Plant 10 tulip bulbs in 10 straight rows so that each row contains exactly 3 tulips.

10. How many cuts would it take to divide a cube of wood into 27 smaller cubes?

11. A cube of cheese is covered on the outside with red wax. It is then cut up as shown below. How many of the smaller cubes have 3 sides covered with wax?

1 side waxed?   2 sides waxed?   4 sides waxed?   no sides waxed?

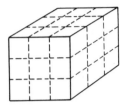

12. Without lifting your pencil from the paper, and without retracing any part of a line, draw no more than four straight lines that will pass through all the dots.

> •   •   •
>
> •   •   •
>
> •   •   •

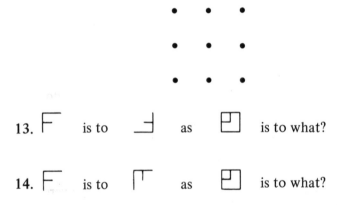

13. ⌐ is to ⌐ as ⌐ is to what?

14. ⌐ is to ⌐ as ⌐ is to what?

15. Two people begin walking side by side. They start so that their left feet hit the ground at the same time. But then one person takes two steps in the same time that the other takes three steps. They continue in this fashion. How long does it take before their right feet hit the ground at the same time?

16. In a certain city, streets with names that begin with a consonant and end with a vowel run north–south. Those streets that begin with a vowel and end with a consonant run east–west. All the rest of the streets run either north–south or east–west.
    a) Willow Street is parallel* to Elm Street. Is Willow parallel or perpendicular to Maple Street?
    b) Cedar Street is perpendicular to Spruce Street. Is Cedar parallel or perpendicular to Oak Street, which runs east–west?

---
*Parallel streets run in the same direction; perpendicular ones meet at right angles.

17. An ant is sitting in the upper corner of a square box with a lid. The box is 1 foot high by 3 feet square. Our ant is heading for the opposite corner at the bottom of the box. This being a very perceptive ant, it chooses the shortest possible route. How far does it travel?

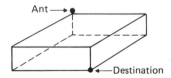

Note: In order to solve this problem one basic mathematical fact is essential. The fact, called the Pythagorean Theorem,* states that the lengths of the sides of every right triangle are related by the equation given below. (*a*, *b*, and *c* represent the lengths of the sides of the triangle as shown and a^2 means $a \times a$.)

Pythagorean theorem:

$$c^2 = a^2 + b^2$$

or $$c = \sqrt{a^2 + b^2}$$

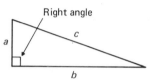

For example, if a = 1 and b = 3, then we find the length of c as follows:

$$c = \sqrt{a^2 + b^2}$$

$$c = \sqrt{1^2 + 3^2}$$

$$c = \sqrt{(1 \times 1) + (3 \times 3)}$$

$$c = \sqrt{1 + 9}$$

$$c = \sqrt{10}$$

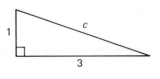

Using a calculator we could find that c is approximately 3.2 feet.

18. A 12-inch ruler is poorly made and is only 11½ inches long. If you measured off what you thought was 2 yards of string with this defective ruler, how much would you actually have?

19. A hunter walks three miles south, then three miles east, then three miles north and ends up where he began. He sees a bear. What color is it?

---

*Pythagoras was a Greek geometor.

20. Make the triangle on the left look like the triangle on the right by moving only three circles.

21. a) In the sketch below remove only two sticks and leave only two squares. There should be no partially formed squares.
   b) Reduce the number of squares to three squares of equal size by repositioning exactly three matches. No matches should be left over.

22. Each figure that follows can be made into a perfect square by making *one straight* cut and repositioning the two pieces. On each figure, sketch where the cut should be made.

a)   b)

c)   d)

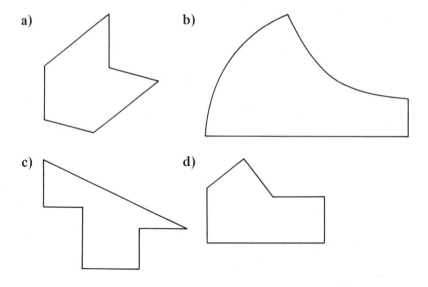

e)

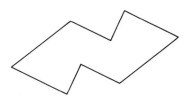

Reasoning Problems

23. How long will it take a one-mile long train to get through a one-mile long tunnel if the train travels at the rate of one mile per minute?

24. Two bicycle riders begin 25 miles apart. Rider A travels toward rider B at the rate of 10 miles per hour. At the same time rider B starts to travel toward rider A at the rate of 15 miles per hour. At the precise time the riders start, a fly begins at rider A and flies toward rider B at the rate of 40 miles per hour. When it reaches rider B it reverses direction without losing any time and flies back toward rider A. It continues to do this until the two bicyclists meet. How far has the fly traveled when they meet?

25. There are 10 stacks of gold bars, with 10 bars in each stack. In 9 of the stacks each bar weighs 1 pound, while in one stack each bar is exactly 1 ounce short of a pound. How could you use an ordinary bathroom scale to determine which stack contains the light bars, if you were allowed to make only *one* weighing?

26. A woman gave a beggar 50¢. The woman is the beggar's sister, but the beggar is not the woman's brother. Who is the beggar?

27. A certain peculiar fellow likes only red, white, or blue socks. He has four pairs of each color, which he throws unpaired into his dresser drawer. If he begins randomly pulling socks from the drawer one dark morning, how many socks would he have to pull to be *sure* of getting a matched pair (two of the same color) from among his collection?

28. Three women enter a room containing five hats, three of them black and two of them white. Without being able to see any hat, each woman picks one and places it on her head. They then file out of the room in single file, facing straight ahead. They cannot see any hats except the ones on the heads of those *ahead* of them. The *last* woman in line says, "I cannot determine what color hat I have on my head." The middle woman says

after hearing that, "I also cannot tell what color hat I have on."
The first woman, who can see no hats at all says, "I *can* tell what
color hat I must have on." What color is it and how does she
figure it out?

29. The natives of Ho Island in the Pacific are of two groups, the
Good Guys who always tell the truth, and the Bad Guys, who
always lie. On a recent visit there, a traveler was approached by
four natives who wanted him to guess to which groups they
belonged. The prize for getting them all right was an ounce of
gold. They stood in a circle around him and offered to answer
one reasonable question, provided he asked each native the same
question. The traveler thought for a minute and then smiled.
He asked each native in turn, "Are you and the native standing
on your left of the same group?" Moving to the left around the
circle they answered: "Yes," "Yes," "No," and "No," Figure out
which group each native belonged to and win the gold.

30. John takes four hours to do a job and Jane takes two hours. If
they work together for one hour, what part of the job will
they finish?

31. If a woman was 29 in 1978, in what year was she born?

32. If a woman was 29 in 1978, how old is she this year?

33. Mary ate 1/3 of a pizza and Jane ate 1/2 of what was left.
   a) How should the remainder be divided so that each gets an
      equal share of the original pizza?

b) If the dog eats the rest, how much should each pay? The pizza cost $3.60.

34. Jim and Mary are both taller than Brad. Sue is between Jim and Mary in height.
    a) Who is the shortest?
    b) Can you determine who is tallest?
    c) Suppose in addition you knew that Sue was taller than Mary. Now who is the tallest?

35. Jim has six times as much volcanic ash at his house as Mary has at hers. Sue has half as much as Brad, and Brad has half as much as Jim. If Mary has two inches of ash, how much do each of the rest have?

36. How many different ways are there to answer a five question true–false test? This is the same as asking how many different solution keys there would be if you were to make up all the possible combinations of answers.

37. Kathy has four times as much money as Joe. Together they have $25. How much does each have?

38. Eustace has four friends whose occupations are butcher, baker, tailor, and carpenter and whose names are Mr. Butcher, Mr. Baker, Mr. Tailor, and Mr. Carpenter, though the names do not correspond to the occupations. Each man has a son and a daughter, but no son practices the same trade as his father and none

practices a trade corresponding to his own name. Each son marries a daughter whose maiden name does not suggest her husband's or his father's trade. Each woman changes her last initial when she marries. The baker's son married Ms. Butcher. Mr. Butcher, Sr. is not a baker. The trade of Mr. Carpenter, Sr. is the same as the maiden name of the woman who married Mr. Butcher, Jr. What are the trades of each of the eight males and the maiden names of the son's wives?

39. The following facts provide sufficient information to answer the questions at the end in only one way.
   a) The Englishman lives in the red house.
   b) The Spaniard owns a dog.
   c) Coffee is drunk in the green house.
   d) The Ukranian drinks tea.
   e) The green house is immediately to the right of the ivory house.
   f) The Tiajuana Gold smoker owns snails.
   g) Kools are smoked in the yellow house.
   h) Milk is drunk in the middle house.
   i) The Norwegian lives in the first house on the left.
   j) The man who smokes Chesterfields lives in the house next to the man with the fox.
   k) Kools are smoked in the house next to the house where the horse is kept.
   l) The Camel smoker drinks orange juice.
   m) The Japanese smokes Lucky Strikes.
   n) The Norwegian lives next to the blue house.

Who drinks Cutty Sark and who owns the zebra?

40. Elden, Mildred, Rita, and Wes are staying one on each floor of a four-story hotel. Their ages are 50, 45, 40, and 25 but not necessarily in that order. Elden is staying directly above the 45 year old and directly below the 40 year old. From her room, Mildred has to pass by the 25 year old to leave the hotel. Mildred is more than one floor away from Wes, who is more than five years younger than Mildred. How old is each person and on what floor are they staying?

9

STUDYING MATH

The following is a discussion of study skills that will be particularly beneficial in math classes. It includes

- *learning styles,*
- *general suggestions for studying math,*
- *how to read math textbooks,*
- *how to take math notes,*
- *suggestions for taking math tests, and*
- *setting math goals.*

LEARNING STYLES

According to some psychologists, our learning style (or cognitive style as it is sometimes called) is the way in which our mind processes information. To others, learning style is also related to our attitudes toward learning and to certain personality characteristics. By recognizing your own learning style, you obtain information that can help you select instructors, courses, and study techniques that work for you.

Identifying learning style is a useful concept because it is nonevaluative. We do not say that one style is superior to another, only different. We each have our own style and the important thing is that we recognize that style and use it to our advantage.

Essential to one learning style model* is the idea that we take in information through the senses—sight, touch, hearing (and in some circumstances, smell and taste). For each of us the relative importance of these senses for learning differs. That is, for one person taking in information through the eyes—whether it is seeing words on a page of a textbook or watching a demonstration—is essential for understanding. For another individual, hearing information is more conducive to learning. Based on this model, a crude way to determine your learning style is to ask yourself, "How do I learn best? Do I learn primarily from reading information, from writing it down, or from hearing it?" In the language of learning styles, you are deciding whether you are a *visual learner*, a *tactile learner,* or an *auditory learner.* Each of these learning styles has implications for choosing classes and study techniques.

visual learner These people may prefer a course that uses a textbook or at least supplies regular, written handouts. They may find courses that are strictly lecture very frustrating. They will find it helpful to have the instructor write information on the blackboard in addition to giving it orally. Slides or films that clarify the material being covered will be useful also. In some courses demonstrations could be important. Outlining and diagraming study notes may be valuable. If you learn best by reading or seeing illustrations you are a visual learner.

tactile learner Writing information down is important for this person's learning. Taking both lecture notes and reading notes may be necessary. Drawing pictures and diagrams may help clarify confusing issues. Classes that offer "hands-on" activities—a chance to touch tools, specimens, materials—should be rewarding for this individual. Although written

---

*Joseph E. Hill. "The Educational Sciences" Bloomfield Hills, Michigan: Oakland Community College, 1971.

homework in math is important for everyone, it may be particularly helpful to tactile learners. If you prefer touching objects and writing down information you are a tactile learner.

auditory learner Hearing lectures is an ideal way for this person to learn. Taping lectures rather than taking notes is most productive. Reviewing course material aloud and studying by talking out loud are useful techniques for the auditory learner. Conversations about the material with other students, either informally or in study groups, will be helpful. If you prefer to listen and you remember what you hear then this is your main learning style.

Even if you do not have a choice of instructors and you cannot match your learning style to an instructor's style you can still apply this information to your method of study. You can bring a tape recorder to class or ask your instructor to draw a picture to illustrate a concept or take notes from your reading, depending on whatever works best for you.

Long before "learning styles" were made popular by educators, most teachers could identify certain patterns in the classroom. They could observe students attitudes toward learning, their social relationship to classmates and the roles they play in the classroom. Five types of learners that emerge from such information observations are as follows.

independent This describes students who prefer to work alone, not requiring the support of classmates or the instructor. They choose what they want to learn and separate what they believe is important from the unimportant. They are confident learners. They often prefer not to initiate classroom discussions even though they may understand the subject well.

uninvolved Students in this group prefer the traditional classroom setting over other learning environments. They are able to remain detached and relatively uninvolved in this setting. They usually do not participate in classroom activities either with instructors or classmates. Often they don't understand what happens in class or they don't care about it (or both). It's hard for them to be interested in school.

social Students in this group enjoy the give and take that comes from sharing their opinions and thoughts with others. They are cooperative and sociable. They use the classroom to form friendships and social connections as well as for formal learning.

receptive This describes students who work hard to learn exactly what they are told and no more. They expect the instructor to have *all* the answers. They want to be given detailed directions that they will then carefully follow. They often rely on the help of classmates outside of the formal classroom.

competitive The primary motive for learning by students in this group is to do better than anyone else. They want to earn top grades and recognition from the instructor and peers.

Each of these approaches to school also has implications for choosing study techniques and classes. The first thing to recognize is that most college level classes are designed primarily with the independent student or the competitive student in mind. Math courses are no exception. This presents potential difficulties for the many students who do not fit into these categories.

If you are a "social" student you may find you learn better when you work in small math study groups outside of class. You may even have to take the initiative in forming such groups since traditionally math teachers have not encouraged group efforts at problem-solving.

If you tend to be "receptive" you will appreciate clear, frequent directions and may have to ask questions to help clarify certain issues, especially if the instructor or textbook are not well organized. You may need to let it be known that you require clear statements of instructor expectations and intentions as well as clarity in course organization and structure.

If you identify yourself as an "uninvolved" student you will probably find classroom instruction less beneficial than do students who fit into the other four groups. But each of us probably changes our learning style several times as we grow and mature. As we are introduced to different instructors or methods of instruction, as we gain confidence, or as our motives for being in school change, we move from being "uninvolved" at one point in our lives or in one situation to being involved at another point. If you see yourself at the "uninvolved" stage you may find it more profitable to leave school until your attitudes or motivations have changed.

Now that you have read about learning styles, let's try to determine the kind of learner you think you are. First, write a paragraph describing how you see yourself as a learner. Then check off the classroom characteristics that would be most helpful to you in completing a math course successfully. Consider such questions as "Where do I sit in the classroom? How often do I speak up? Do I like to ask questions? Answer questions? Do I prefer working alone or in groups? Do I take risks and pursue the topics that interest me the most or do I generally follow what the instructor or textbook suggests?

my learning style _____

_____

_____

_____

_____

Exercise: COURSE CHARACTERISTICS CHECKLIST

Directions: Place a check beside any of the following that appeal to you as a learner.

_____ Small class (15 or fewer)

_____ Large class (more than 15)

_____ Lecture class

_____ Small group discussion and problem solving activities (3 to 5 students)

_____ Independent study—using a self-study textbook and working on your own

_____ Independent study—exploring a topic of your choice on your own; choosing your own reading material; seeing the instructor only to report on progress and to get assistance

_____ Computer-directed learning*

_____ Combination of lecture and group problem solving sessions

_____ Individual tutoring

Look over your checks and use the following space to describe the learning situation that fits best with your learning style.

_____

_____

_____

_____

_____

_____

Now that you have a picture of yourself as a learner, you may be better able to select the math learning situation that is best for you. Sometimes, however, there is little choice; in any case it is important to master the study skills that apply directly to math classes.

GENERAL STUDY TIPS

- Sit in the front of the classroom. This will not increase significantly your chances of being called upon by the instructor. It will ensure that you can see and hear the material being covered. And it will help counter any tendency you may have to "tune out" the instructor or to daydream.

- Bring the tools of a math student to class *each time*. These always include plenty of paper and pen or pencil for notetaking and working on problems. (Try to use pencil on tests unless ink is required—and keep a large eraser handy.) Other useful items often include course textbook, calculator, special graph paper, or a ruler. Don't force yourself to work under a handicap by being unprepared.

---

*This involves working alone at a computer terminal, reading explanations and questions on a TV-type screen, typing in your answers, and learning from the computer whether or not your answers are correct.

Bring the tools of a
math student to class
each time . . .

- Ask questions if you don't understand. If at first you are not comfortable speaking out in class, find out the instructor's office hours and ask him or her in the office.

- Use flash cards as a memory aid. If you have to memorize such things as vocabulary or formulas, write the word or title of the formula on the front of a 3 × 5 card and the definition or formula on the back. Go through your cards saying the word or title out loud. Then give the definition or formula. Turn the card over and see whether you were correct. Put the cards for which you gave a correct answer in one pile and the other cards in another pile. Work on the error pile again until you have mastered it. Also recheck the "correct pile" occasionally.

- Try not to "tune out." If you are feeling anxious or confused, it is better to recognize and accept those feelings than to try to avoid them. If you "tune out" and don't listen to what is being said in class, it won't help the anxiety and you'll probably miss key information (making you more anxious). Instead try using positive self-talk to acknowledge the feelings and take notes anyway as best you can.

- Try to relax. You can't swim when your muscles are tense and you can't do math when you're feeling panicky.

- Attend class everyday. Math classes are hierarchical. That is, what you learn today will be necessary for understanding tomorrow's material.

- Do homework regularly (whether assigned by the instructor or not). Recognize homework as your chance to practice without being penalized for making errors. You'll probably find that setting aside a specific time each day in a nondistracting environment works best. If your textbook has answers to problems check your answers. When your answers do not match those in the book jot down some questions to ask the instructor the next day. Don't spend a lot of time worrying about why they don't match.

- Don't get behind. This is probably the most important suggestion we can make. Falling behind will increase your anxiety tremendously and it is almost impossible to play "catch up" in a math class. It's not like falling behind in your reading for a sociology course. You will not be able to learn the material at the last minute by putting in extra hours all in a row. It takes time to feel comfortable with math and to have it sink in. Also, you generally need time to master one step before you can fully appreciate the next.

MATH STUDY SCHEDULE

		1 Chapters 1+2	2	3	4 Ch. 3 →	5 more Ch.3 !
6	7 Ch.4	8 Chapter 5	9 CH.6	2 10 call Bob re: ch 5+6	11 QUIZ Ch. 1-6	12 Ann's 7:30p.m.
13 Chapter 7	14 ← Chapters 8-10 →	15	16	? 17 call Gran w: pg. 88	18 Review for TEST Monday	19
20	21 PRE TEST	22 Review	23	24 To BED EARLY	25 FINAL EXAM	26 Movie w/ Sylvia
27 Dinner at Mom's	28	29	30	31		

READING MATH TEXTBOOKS

First you must recognize that math textbooks are not novels. The speed-reading techniques you may have learned will probably not be helpful when reading mathematics. Often every word in a math sentence is important. Unlike novels, you cannot skim math books quickly and still understand the main points. The latter must be read slowly, carefully and many paragraphs must be reread several times.

Second, math textbooks often use words that we read and use everyday. However, these words sometimes have different meanings from our normal usage. Therefore it is important to study definitions carefully and to learn the *mathematical* meanings of these words.

Third, math textbooks often give examples to illustrate how to solve particular kinds of problems. These worked-out examples are meant to guide you but they often leave out intermediate steps. It is expected that you will read with pencil and paper in hand and be sure to fill in enough additional steps to make the example clear to you. You may want to ask your instructor for assistance on some of the more difficult examples. Understanding examples is the key to understanding the material. Also read any graphs or charts you encounter. These could be important in untangling a concept or problem.

Finally, don't forget to look for the main points, the key ideas in your effort to understand the details. Pay attention to chapter headings and subheadings as these will provide clues to understanding the "big picture." Consider outlining or summarizing your reading as you go along.

TAKING NOTES

The purpose of notetaking is to help you clarify and remember what you read and hear. Notes are only for you, so the system you develop does not need to be intelligible to anyone else. What matters is that it is intelligible to you. You may want to write word for word what you hear in a lecture, but this is generally impossible. You may want to develop your own shorthand—a system of abbreviations and symbols that allows you to write information down quickly. You may prefer to take notes in the form of pictures or diagrams that show relationships between ideas. You may want to recopy notes about certain important topics onto separate index cards for reference. An example of a card on solving linear equations might look like the following. If your algebra is rusty, don't worry about understanding the details on this one.

Linear Equations (one variable)

(also called first-degree equations)

Form: $ax + b = 0$ (a, b real numbers and $a \neq 0$)

Example: $3x - 8 = 6(x + 1) - x$

Solution:	$3x - 8 = 6x + 6 - x$	— clear brackets
Method	$3x - 8 = 5x + 6$	— combine terms
	$3x - 8 - 5x = 5x + 6 - 5x$	— "x" terms to one side
	$-2x - 8 = 6$	— combine terms
	$-2x - 8 + 8 = 6 + 8$	— constants to other side
	$-2x = 14$	— combine terms
	$x = -7$	— divide by coefficient of x

Regardless of your personal note-taking system, there are certain tips that are useful for taking notes in a math class.

1. Every symbol in a formula has a specific meaning. You cannot omit pieces when copying equations or formulas. You also need to copy accurately. For example lower-case "d" and upper-case "D" are considered different symbols in mathematics.

2. Review your notes daily, preferably right after class. Make sure you understand them and fill in any missing details while you can still remember what was said. Ask the instructor for help the following day if you have missed getting something significant into your notes or don't understand something you've written.

3. Use your notes as a supplement to the text. Try to see the relationship between the two rather than looking at them as totally separate sources of information.

4. Take additional notes while reading the textbook. This helps to reinforce ideas in your mind.

5. From time to time go back and outline what you have studied so that you can see relationships between various topics covered.

6. Keep a list of all the important vocabulary words as you go along. You may also want to keep a list of important principles (called theorems). These are helpful when you are reviewing for an exam.

TEST-TAKING TIPS

You need to apply test-taking skills before, during, and after a math test. Before the test the emphasis is on preparation. If you have been doing all assigned reading and problems, you are halfway there. Ideally, you begin to prepare specifically for the test at least one week in advance. During this time you may review your old homework assignments, classroom notes, and the key ideas from the textbook. You should rework several examples of each kind of problem even if you think you already know how to do them. This is the time to doublecheck.

You also need to step back and review the general concepts and main ideas that have been presented. It is easy to overlook these larger issues in the rush to do the problems. However, seeing patterns to the material and being able to generalize may be critical to real understanding and to ultimate success in the course. It is fair to ask specific information of the instructor about what will be covered on the exam. What does he or she feel is most important? What will be emphasized? Sometimes instructors provide samples of their previous tests for reference. If so, be sure you study them for clues as to what will be asked.

Continue to review just *past* the point where you think you really know the material. Go a step further and "over-learn" it. At this stage you will feel saturated. You have done more than you felt you needed. You have done several extra problems of each kind; you have practiced definitions to the point you are dreaming of them; you have reviewed your notes at least two extra times beyond the time you believed you understood them fully. This kind of thorough preparation will go far to relieve anxiety at the time of the test.

The night before the test let studying math be the last of your tasks. This ensures that you do not encounter "interference" from other subjects. Then go to bed and rest well. You'll make careless mistakes and experience blocks if you try to do the test on very little sleep. Your preparation will make extra cramming at this time unnecessary.

On the day of the test be sure to take an adequate supply of the necessary tools—paper, pencils with erasers, calculators (with good batteries) etc. (Avoid using pen on exams; small mistakes become big time wasters.) The time just before the test will be best used in putting yourself in the right state of mind rather than doing further studying. You can increase your anxiety by talking to classmates, especially if you are comparing yourself to them at the same time. "She's smarter than I am." "He studied longer." "She studied different material." Your fear will also be increased by trying to remember some formula you suddenly forgot and by shuffling through your notes or textbook for some last piece of information. Rather than doing any of these activities, empty your mind of distracting thoughts, especially the unpleasant ones. Take several deep breaths; shut your eyes; imagine a pleasant scene. Repeat your positive self-statements. Work on putting yourself in a positive, calm frame of mind. This is the time to use the relaxation techniques discussed in Chapter 5.

When you first receive the test, look it over briefly before starting in to work. Read all directions *carefully*. Determine which questions count the most (you have a right to ask if it isn't indicated). Then proceed through the test taking each problem as it comes. If you get really stuck, move on to the next question fairly quickly. Return to unfinished work if time permits. Try to save some extra time for those questions that carry the greatest number of points.

If you have memorized formulas for the test, quickly jot them down right at the beginning, so you do not have to worry about remembering them later. If you feel very nervous, begin with a question that is easy for you to answer. Your confidence will build as you have a few successes. You will also get your mind in gear and may pick up information that is helpful in other problems. Make sure you do exactly what each question asks you to do. Be alert for phrases or words such as "reduce to," "solve for," "define," "describe," "prove," and "leave the answer in (some special form)." Each of these is directing you to perform a specific action. Many students allow their anxiety to interfere with their understanding of these directions. Generally you will not receive credit if you perform some action other than the one you were directed to do, even if you do the problem correctly. Correct work, which does not answer the question asked, is a waste of your time. For the same reason, try not to do work beyond what is required.

If you find yourself becoming excessively nervous or blanking out during the test, stop for a short time. Close your eyes, breathe deeply, repeat your positive self-statements. When you feel more composed return to the test. The time you lose taking this short break is more than compensated for by the improvement in your performance after the break. Usually this takes less than a minute although it may seem much longer. If you are never able to perform well in a timed test and you think this is due to test jitters, discuss this with your instructor. Perhaps your school has a testing office and arrangements can be made for you to take the test there without time constraints. Occasionally an instructor will allow you to demonstrate your knowledge and understanding in some other form. In any case you will have made her or him aware of your problem, which may in itself be helpful.

Proper test taking does not end when the test is over. When you receive your corrected test back, go over the entire thing. Correct your errors, with the instructor's assistance if necessary. Be sure to ask about any corrections you don't understand or that you disagree with. Often discussing that item with the instructor will be just the thing that finally makes it come clear. If you have not done as well on the test as you hoped you might ask the instructor for suggestions on how to study

differently for the next one. Finally, when studying for the next exam be sure to come back and review this one. Go through it again, reworking all the problems, not just the ones you missed.

SETTING MATH GOALS

Now is the time to establish some math goals while the information you have read is still fresh in your mind and your confidence somewhat improved.

Defining and accomplishing math goals, like any other goals, is best done according to a system and following certain rules. These rules are as follows.

1. Choose your own goals. Pursue mathematics because *you* want to gain mastery over some aspects of the subject or because *you* wish to enroll in a field of study that requires it, not because someone else thinks it would be good for you. Expect that your goals may change. Re-evaluate them and make changes that make sense to you as time goes on.

2. Set goals you can complete. If you do not remember any math since the third grade, then setting an initial goal of completing

calculus immediately is unrealistic and self-defeating. Your first goal may be to learn how to work with fractions, your second to understand algebra. Completing a calculus course would be appropriate as a long-range goal. Don't be embarrassed to start easy but don't underestimate your ability either. You want your goal to be achievable but you also want it to be challenging so you will feel satisfaction upon reaching it.

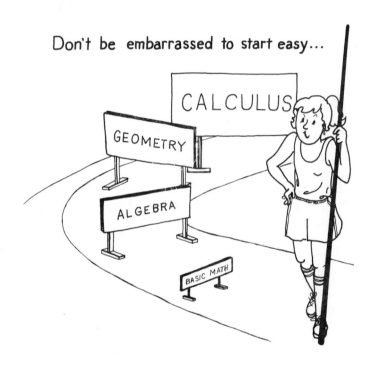

Don't be embarrassed to start easy...

3. Be definite about your goal, so it will be clear when you have reached it. For example, don't say, "In three weeks I'm going to get better in math." Instead your goal would be something like, "In three weeks I will be able to multiply and divide whole numbers." This is a very specific goal and you will easily recognize when it has been completed.

4. Identify people close to you who will support you in reaching your goal. Let them know what difficulties you expect and how they can help.

5. Reward yourself. It studying math is stressful and gives you little instrinsic pleasure, find a way to treat yourself well as you work on it. For example, if you find it hard to stick with your resolve to do math homework every night, think of some small activity you really enjoy. It could be a short walk, calling a friend, reading a magazine, taking a warm bath, getting a massage, etc. Then, when you've put in your time on the homework, take the time to reward yourself. Or, if you've been studying hard for a math test, afterwards allow yourself to go to that movie you've been wanting to see for months.

6. Break your goal into small steps that tell you exactly what you have to do.

Example Goal: Complete Math 105 next quarter with a C grade or higher.

Steps: 1. Fill out college admissions application form.

2. Arrange to see advisor.

3. Pick up copy of previous transcript from the university to show that I have already completed Math 101.

4. Find out which teachers best match my learning style.

5. Register for Math 105.

6. Get book before school starts; preview it.

7. Inquire about tutoring, just in case I need help.

8. Set up study—work—play schedule; be realistic! (then follow it).

What is your current math goal? Try to pick one that can be completed in three to six months.

Exercise: MY MATH GOAL

Directions: Write down your present math goal. Then list the steps you need to take to complete it, the dates you will complete each step, ways in which you will reward yourself, and names of people who can assist you.

Target Date for Completion of Step

Small Step: 1. _____   _____

2. _____   _____

3. _____   _____

4. _____   _____

5. _____   _____

6. _____   _____

Possible Rewards: _____

When I need more support in reaching this goal I know I can call on:

_____
person's name

_____
person's name

_____
person's name

FURTHER READING

Frand, Jason, L. *How to Study Mathematics, Chemistry, Statistics, Physics.* Culver City, Calif: SKIL Publishing Co., 1979.

Galarosa, Annie, Ann Oxrieder, and Marilyn Weckwerth. "How to Take Tests." Seattle, Wash. Seattle Central Community College, 1975.

Oxrieder, Ann, Mike Linscott, Sanford Helt, and Rafael Marino. "How to Read a Math Textbook." Seattle, Wash. Seattle Central Community College, 1978.

Shaw, Harry. *Thirty Ways to Improve Your Grades.* New York: McGraw-Hill Book Co., 1976.

"The Key Word Approach to Note Taking," filmstrip available in Student Assistance Center, Seattle Central Community College, 1701 Broadway, Seattle, WA 98122.

CALCULATORS
AND
CONSUMERS

In this chapter we will
- *briefly explore the mysteries of calculator and operation, and*
- *discuss the use of calculators in a few commonly encountered daily situations.*

INTRODUCTION

The average person encounters mathematics, in some form, several times in the course of a week. The exposure may be something as trivial as deciding a proper tip at a restaurant. However, it may involve a significant decision, like how to finance a large purchase. In many cases, the mathematical questions that confront consumers can be answered using only elementary arithmetic and a few simple formulas. Yet many of us choose to leave the calculations for others to perform, when we could easily learn to do them ourselves. We are therefore at the mercy of the people and organizations with which we deal. Even if they are all very honest, it is always possible that *they* will make a mathematical mistake. It is certainly in our own best interest to possess the confidence and ability to at least check someone else's figures.

Often the calculations necessary involve only basic addition, subtraction, multiplication, and division, but the details may be time consuming and boring. If the operations had to be performed longhand, it wouldn't be worth the effort. Fortunately we live in an age where inexpensive aides to calculation are readily available. Small, hand-held calculators can be purchased for about the same cost as a good restaurant meal. We believe that some type of simple calculator is a good investment for anyone. Math-anxious individuals particularly benefit from owning one, learning how to use it properly, and then practicing with it frequently. This has two advantages: (1) when you use a calculator to check someone else's numbers, you protect your interests as a consumer; and (2) by forcing yourself to deal with numbers on a regular basis you increase both your confidence and your skills.

It is surprising how many people feel that using a calculator is an admission of incompetence or maybe even some form of cheating. After all, shouldn't you really be able to do it by hand, or in your head? Embarrassment prohibits some people from using a calculator in public, even though they may in fact own one and have it available. The following true story may illustrate how self-conscious we can become. A group of women went to lunch, each intending to pay for their own share of the meal. The restaurant would not provide separate checks so when the bill came there was the usual general confusion of trying to decide who owed what. Finally one person in the group took out a pencil, paper, and calculator, then proceeded to list each person's purchases and find their individual totals. This is not terribly remarkable except that the person who used the calculator happened to be a mathematician. One would think she ought to be able to do the addition by hand. Perhaps she could, although math people are not necessarily good at addition. In this situation, however, to do mental arithmetic was not necessary. Using the calculator was faster and more accurate. After leaving the restaurant, another woman in the group confided that she too was carrying a calculator, but was too embarrassed to use it to figure out the bill.

SAMPLE QUESTIONS

The following is a sampling of problems of the type each of us probably encounters from time to time. The methods of solution will be described in detail in later sections. Although all of them *can* be worked using only pencil and paper, you are encouraged to use a calculator. You will need only the simplest type, one that performs the operations of +, -, ×, and ÷. If you are thinking of buying a calculator, you may want to consider one with a few more features so that as your math ability increases and your needs become more complex, your calculator will be adequate to handle the task. If you decide to buy one, it is a good idea to discuss your requirements with someone at a store that carries several types. They should be able to help you decide on one that's right for your needs and your pocketbook. To begin with, we strongly suggest an "arithmetic" calculator rather than an "RPN". (RPN stands for Reverse Polish Notation.) These two designations refer to the method of entering numbers and giving instructions to the calculator. The "arithmetic" type operates in the same way that we usually do arithmetic. Its operation therefore seems more natural and is easier to understand. No matter which you select, you will need to study the instruction book that comes with the calculator carefully. Each instrument does things just a little bit differently.

The following situations are typical ones in which a calculator can be used to advantage. With the knowledge of a few simple principles and a tool to do the arithmetic, you can easily complete the necessary computations. First we will go over the basics of calculator use and then return to solve these problems.

1. You find an item on sale for 30% off. How do you calculate the proper reduced price?

2. You purchase a meal at a restaurant and need to decide on how much to leave for the tip.

3. You want to figure out the mileage you're getting on your old car.

4. After computing your mileage and checking the price of gas, you decide you had better buy another car that gets more miles per

gallon. You can borrow money to purchase the car from either of two sources, which quote you different interest rates. Which loan will cost you the least?

5. Aunt Florence just died leaving you a small inheritance. How do you decide the best place to invest the money based on the interest rates you are given?

6. Your monthly charge card statement comes in and you think it contains a mistake. How do you check the figures?

CALCULATOR OPERATION–GENERAL GUIDELINES

The directions in this and later sections are for an "arithmetic" style calculator. We begin by discussing the basic keys on the calculator and then give a few practice problems. (In some cases you may be able to shorten the process described by reading directions for your own calculator.) The following general comments will be more understandable if you read them with a calculator in front of you.

The Basic Keys

- **ON/OFF**: The calculator will either have an on/off switch or an on/off key. To begin a problem turn the power on; usually the display portion will light up and show 0.

- **MODE**: Some machines perform several different kinds of calculations. Read your own instruction booklet to find out if this is the case for yours. If so, be sure to select the proper mode for basic arithmetic.

- **OPERATIONS**: You will have keys with the four operations of arithmetic on them: ⊞ ⊟ ⊠ ⊡

- **NUMERALS**: There will be a key for each of the numerals 0 through 9.

- **DECIMAL POINT**: This key, ⊡ , inserts a decimal point in a number. For example 23.1 has a decimal point between the 3 and the 1.

- **CLEAR:** This key erases *all* previous entries from the calculator. (If your machine has a memory, the clear key will not erase the memory.) The clear key may be labeled: [c]   or   [AC]   or you may just press the   [ON]   key twice.

- **CORRECT:** This key erases only the *last* number you have entered in the calculator. Everything done before that entry remains unchanged. This is useful if you make an entry error in the middle of a problem. The error correction key may be [CE]   or you may press the   [ON]   key just once.

- **EQUALS:** When the equals keys, [=] , is pressed, the calculator displays the end result of all computations you have done so far. It can be useful both at the end or in the middle of a series of operations.

SAMPLE CALCULATIONS

In each of the examples below, be sure to clear the calculator using the Clear key before beginning the next problem. Numbers are entered by pressing the appropriate keys reading from the *left* side of the number.

example 1: Enter: 3,407
    Press in order: [3] [4] [0] [7]

example 2: Enter: 3,407.69
      Press in order: [3] [4] [0] [7] [•] [6] [9]
    Notice that commas written in a number are ignored. Now try each of the examples below. Enter the numbers as illustrated in the examples above. Only the operation keys are shown separately in the remaining examples.

example 3: Find: 301 + 78
    Press: 301 [+] 78 [=]
    Result: 379

example 4: Find: 5,678 + 29.7 + .005
    Press: 5678 [+] 29.7 [+] .005 [=]
    Result: 5707.705

example 5: Find: 689 ÷ 13
Press: 689 ⌈÷⌋ 13 ⌈=⌋
Result: 53

example 6: Find: 689 ÷ 14
Press: 689 ⌈÷⌋ 14 ⌈=⌋
Result: 49.214285

The numbers in Example 6 are very close to those in Example 5. So why are the results so different looking? First notice that 53 and 49.214285 are not really far apart in size, since the latter is approximately 49. The calculator is telling us that 13 divides evenly* into 689, but 14 does not. When a division is encountered that is not an even one, the calculator simply shows the approximate answer to as many decimal places as will fit in the display. You may not be interested in the entire number shown. In some applications you may want to "round off" the result in some way. To round off to the nearest whole number you examine the number immediately to the right of the decimal place. If this number is 4 or smaller (0, 1, 2, 3, 4) then just remove the decimal part (all the numbers to the right of the decimal point). If the number just to the right of the decimal point is 5 or greater (5, 6, 7, 8, 9) then you must increase the number just to the *left* of the decimal place by one unit and remove the decimal part. For example, 23.2701 rounded to the nearest whole number is 23; but 23.6913 rounded to the nearest whole number is 24.

example 7: Find: 61 ÷ 3 and round to the nearest whole number.
Press: 61 ⌈÷⌋ 3 ⌈=⌋
Result: The display will show 20.333333. Rounded to the nearest whole number, the answer is 20.

In the next few examples, there is more than one operation per problem. In such cases, you must decide which operation should be done first. There are several rules from arithmetic that guide us:

1. Any operations appearing inside brackets, ( ), should be calculated first. In (2 + 3) × 5, you first add 2 + 3, and then multiply by 5. The result is 25.

2. If no brackets are shown, then do multiplication and division *before* doing addition or subtraction. So in 2 + 3 × 5, you would first multiply 3 × 5 and then add 2. The result is 17. Compare this to the result in (1) above.

---

*A division is said to be "even" when the result yields a whole number, ones like 1, 2, 3, 4, 25, 31, and so on.

3. If neither rule above applies, then do the operations in order from left to right. For example, 2 – 3 + 4 should be calculated as 2 – 3, then add 4. The result is 3.

example 8: Find: (123 + 45) × 5
Press: 123 ⌊ + ⌋ 45 ⌊ = ⌋   ⌊ × ⌋ 5 ⌊ = ⌋
Result: 840

We pressed "equals," in the middle of the problem so that the calculator would finish finding the value of 123 + 45 before it multiplied by 5. After that we multiplied by 5. This was because 123 + 45 was in brackets and therefore should be done first.

example 9: Find: 123 + (45 × 5)
Press: 45 ⌊ × ⌋ 5 ⌊ = ⌋   ⌊ + ⌋ 123 ⌊ = ⌋
Result: 348

In this case we had to begin at the end of the problem, (multiplying 45 × 5), because this operation was in brackets. You should notice that in Examples 8 and 9 the numbers and the operations are identical. The only thing that distinguishes them is the location of the brackets. Since the results are not equal, you can see how important it is to watch for brackets.

example 10: Find 24 + 6 and 6 + 24 and compare results.
To find 24 + 6: Press 24 ⌊ = ⌋ 6 ⌊ = ⌋ . The result is 30.
To find 6 + 24: Press 6 ⌊ + ⌋ 24 ⌊ = ⌋ . The result is 30.
The results are equal.

example 11: Find: 24 ÷ 6 and 6 ÷ 24 and compare results.
To find 24 ÷ 6: Press 24 ⌊ ÷ ⌋ 6 ⌊ = ⌋   The result is 4.
To find 6 ÷ 24: Press 6 ⌊ ÷ ⌋ 24 ⌊ = ⌋   The result is .25
The results are not equal.

These two examples illustrate the importance of order in performing operations. For addition, 24 + 6 is the same as 6 + 24. But for division 24 ÷ 6 is unequal to 6 ÷ 24. Use your calculator to decide if 24 – 6 equals 6 – 24. Then check to see if 24 × 6 equals 6 × 24. You will notice that for addition and multiplication the order of doing the operation doesn't matter. But for division and subtraction *the results obtained are different*. In the next example, notice that we are careful to perform the division in the proper order.

example 12: Find: (68 + 4) ÷ (15 – 3)
Press: 15 ⌊ – ⌋ 3 ⌈ = ⌉
Record the result, which is 12, on a piece of paper and then clear the calculator.

Press: 68 $\boxed{+}$ 4 $\boxed{=}$ $\boxed{\div}$ 12 $\boxed{=}$
Result: 6

In this problem we first computed 15 − 3 and kept track of the result, 12. Then we completed 68 + 4 and finally divided by 12. On some calculators there is a built in memory that will keep track of an intermediate result like 12. Can you decide why we chose to perform the operation inside the second brackets first? Try to do this same problem in some other order and see if you can come up with the same result.

example 13: Find: 4 × [3 + (62 ÷ 5)]
Press: 62 $\boxed{\div}$ 5 $\boxed{=}$ $\boxed{+}$ 3 $\boxed{=}$ $\boxed{\times}$ 4 $\boxed{=}$
Result: 61.6

When more than one set of brackets is "nested" inside another, as in Example 13, we begin at the innermost brackets and work our way out. Now try to put all these facts together in the next example.

example 14: Find: [(13 − 3) × 7] ÷ (34 − 29)
Try to compute the result with your calculator before checking with the solution below. The correct value is 14.
Solution: Compute 34 $\boxed{-}$ 29 $\boxed{=}$ and record the result (5)
Compute 13 $\boxed{-}$ 3 $\boxed{=}$ $\boxed{\times}$ 7 $\boxed{=}$ $\boxed{\div}$ 5 $\boxed{=}$
Try the following two examples just for fun.

example 15: Compute 2886.7 × 20, turn the calculator upside down, and read.
Then *add* 4 and read again.
Result: HELLS BELLS

example 16: If $28,430,938 worth of oil is sold at a profit of 2.5%, find the value of the profit. (How to find some percent of a number will be discussed in more detail later in this chapter.)
Press: 28430938 $\boxed{\times}$ 2.5 $\boxed{\times}$ .01 $\boxed{=}$
Result: $710,773.45 (and turn the calculator upside down to see who receives the profit.)

SAMPLE QUESTIONS REVISITED

We are now ready to work on some of the specific questions mentioned before. In some cases a formula from mathematics or business will be used in the course of the solution. You are *not* expected to know that formula in advance or to decide where it came from. Although it is interesting and often helpful to know how a formula is developed, that can be the subject of an entire course. There is an important lesson to be learned here. We are perfectly capable of *applying* a formula to solve a problem even when we do not fully understand how the formula was obtained. The ability to insert the correct numbers at the appropriate places and a certain trust or belief in the validity of the formula is all that is required. Of course to fully understand the concepts covered, a study of the mathematical principles involved is necessary.

Question 1

You are considering buying a stereo component that is on sale. It normally sells for $389.95 and is now advertised for 30% off. How much will you have to pay?

solution: First we can find out how much is to be taken off from the original price ($389.95). Then we must subtract this figure from the original price. A formal discussion of the meaning of percent and the process of converting from percent to decimal form is not of interest to use now, although it is an important topic in arithmetic. (See the references at the end of this chapter for more information.) Here we will merely show a simple method to compute a certain percent of a given number. 30% is entered into the calculator by entering 30 and then multiplying by .01. That is, 30% is the same as 30 × .01; similarly 5% is the same as 5 × .01 and so on. So to find 30% of $389.95 we calculate:

$$30 \times .01 \times \$389.95 = \$116.985$$
$$\underbrace{}_{30\%}$$

Notice that the word "of" in 30% of $389.95, translates into "multiply" when we do the calculations. The result, $116.985, is the dollar amount of the discount. To find the sale price, we subtract this from the original.

$$
\begin{aligned}
\text{Sale price} &= \text{Original price} - \text{Discount} \\
&= \$389.95 \quad\quad - \$116.985 \\
\text{Sale price} &= \$272.965
\end{aligned}
$$

At this point you may want to round off this number to the nearest penny (two places to the right of the decimal point). So your actual cost would be $272.97.

Question 2:

Your bill at a restaurant comes to $27.83, before tax is added. The service was adequate, but not exceptional, and you would like to leave a tip of 15%. How much is the tip?

solution: We want to compute 15% of given total, that is, 15% of $27.83. As in the previous question we do the following:

$$15\% \text{ of } \$27.83 = 15 \times .01 \times \$27.83$$

Using a calculator, we get $4.1745, which rounded to the nearest penny is $4.17. That is the tip.

alternate solution: In many states the tax on a meal is about 5% of the cost of the meal. Since 15% is 3 times 5%, you can find the approximate value of the tip by multiplying the tax by 3. In the example above a typical bill might look like:

Food, etc.	$27.83
Tax (at 5.2%)	1.45
Total bill	$29.28

The tip would be about 3 times the tax, or 3 × $1.45 = $4.35. This is close to the result we obtained previously.

Question 3:

When you filled your gas tank the time before last, the speedometer read 43,276.8. This time when you filled your tank it read 43,510.4. You add a total of 19.5 gallons to fill the car. What is your mileage (miles per gallon) on this tank?

solution: First we need to compute the total miles traveled. Subtract your previous speedometer reading from the current one.

Miles traveled = Current reading − Previous reading
Miles traveled = 43510.4   − 43276.8   = 233.6 (miles)

So you have used 19.5 gallons to go 233.6 miles. Can you figure out what to do with these numbers to get miles per gallon (mpg)? Let's suppose you're not sure. How could you decide what to do? One strategy would be to simply try the four operations of arithmetic on these two numbers and decide if any of the answers are reasonable. We now show the results of the eight possibilities. Which, if any, seem to be sensible?

233.6 + 19.5 = 253.1	or	19.5 + 233.6 = 253.1
233.6 - 19.5 = 214.1	or	19.5 - 233.6 = -214.1
233.6 X 19.5 = 4555.2	or	19.5 X 233.6 = 4555.2
233.6 ÷ 19.5 = 11.979487	or	19.5 ÷ 233.6 = .083476

About the only result that is reasonable for mileage is 11.979487; that is 233.6 ÷ 19.5 or (Number miles driven) ÷ (Number of gallons). Do you see how just trying all the possibilities and using your good common sense can solve a problem? We might also want to try to generalize this, and say that:

$$\text{Miles per gallon} = (\text{Miles driven}) \div (\text{Number of gallons})$$

Question 4

You are planning to buy a used car that costs $3,500. You want to pay $500 in cash and take out a loan for the remainder of the purchase price. The car dealer's plan involves paying off the balance at $142 per month for the next two years. How much interest will you have paid at the end of this time?

solution: The interest is the price you pay to borrow the money for the balance. The amount you pay that is above the actual amount borrowed ($3000) will be the interest. First let's determine the total amount that will be paid back to the dealer on the $3000 loan.

$$\text{Total payments} = (\text{Monthly payment}) \times (\text{Number of months})$$
$$= \$142 \times (12 \times 2)$$
$$\text{Total payments} = \$3408$$

Now subtract the amount borrowed from how much you actually paid back.

$$\text{Interest} = (\text{Total payments}) - (\text{Amount borrowed})$$
$$= \$3408 - \$3000$$
$$\text{Interest} = \$408$$

So you paid $408 in interest.

You may recall that interest rates on loans are most often given in terms of percentages, like a loan at 10%, or 15%. However there are many different methods of computing interest*. What may sound like a lower percentage rate loan may actually end up costing you more. For example you may actually pay more in interest on a 12% loan computed one way than on a 13% loan computed another way. One way of comparing loans however, is by calculating the total amount of interest paid over the life of the loan. For example compare the cost of the loan described below, with the loan offered by the car dealer.

Example: You find that a finance company will also loan you the $3000 to buy the car. They offer you a three-year loan to be paid back at $103 per month. Compute the total amount paid in interest.

$$\text{Total payments} = (\text{Monthly payment}) \times (\text{Number of months})$$
$$= \$103 \times (12 \times 3)$$
Total payments = $3708
Interest   = $3708 – $3000
Interest   = $708

You will pay $708 in interest over the three-year life of this loan, compared with $408 in interest on the two-year car dealer's loan. In some cases of course you may deliberately choose the more expensive finance company loan in order to have lower monthly payments. Another way to compare the relative costs of two loans is to ask what the "true annual interest rate" would be. This is often not the same percentage as the loan rate you are quoted. True annual interest rates can be computed using a mathematical formula or from tables available at most local banks. We will not discuss them here.

Question 5

Interest rates are also involved when you wish to invest money. Suppose you had $1000 to invest for a period of at least five years. There are two options available to you. One bank will pay 6% interest compounded annually while another will pay you 6% simple interest. Which type of account will yield the best return?

solution: Since the interest rates are the same we might guess that the rates of return would be the same in both accounts. In the case of compound interest, however, each time interest is compounded, that money is added to your account and you then earn interest on a larger amount for the next period. In the case of simple interest, you earn money only

---

*You may be charged interest for the entire period of the loan on the total amount borrowed; the interest may be charged only on the unpaid balance; and so on.

on your original investment, no matter how long it is left on deposit. Since you are paid interest on a larger amount in the compound account, it seems reasonable to suppose the return would be greater on this account.

We have intuitively answered the question of which is a better deal, but suppose we also wanted to know exactly how much money would be earned in each account? We can compute these figures with the help of two formulas from business. We will begin with the simple interest account. Basically, the interest you earn is the product of how much you invest (called the principal), multiplied by the annual interest rate (expressed as a decimal), multiplied by the number of years the money is left on deposit. In our case, the principal is $1000, the interest rate is $6\% = 6 \times .01 = .06^*$, and the number of years is 5. So the computations look like:

$$
\begin{aligned}
\text{Interest} &= \text{(Principal)} \times \text{(Interest rate)} \times \text{(Years)} \\
&= \$1000 \quad\quad\quad \times .06 \quad\quad\quad\quad \times 5 \\
\text{Interest} &= \$300
\end{aligned}
$$

So we would receive $300 in interest at the end of five years in the simple interest account.

The formulas for compound interest are derived from those for simple interest in a fairly straightforward way, but it is not our intention to show that here. In the case of compound interest, it is easier to calculate the total amount on deposit at the end of the time period first. We then find the amount of interest earned by subtracting the original principal from the total. The formula for the total amount is shown below.

$$
A = P \left(1 + \frac{r}{n} \right)^m \qquad \text{where} \qquad m = t \times n
$$

Does this look totally incomprehensible? If so, that is natural since you can't expect to understand what is intended until you know the meaning of the symbols involved. You must first find out what each of the letters stands for, and how to perform all of the operations. Mathematical formulas are often given in letters, where each letter represents some quantity. The value of that quantity in any application is determined by the given information. So first we will define the meaning of each letter.

A represents the total amount on deposit at the end of the given time.

P is the principal (the amount that is initially deposited)

---

*Recall from Question 1, to convert percent figures to decimals on the calculator, you multiply the percent by .01.

r stands for the (annual) interest rate, expressed as a decimal
n is the number of times a year the interest is compounded
t is the number of years the money is left on deposit

In our example, A is the value we want to compute, $P = \$1000$, $r = 6\% = 6 \times .01 = .06$, $n = 1$ (since interest is compounded annually, meaning once a year), and $t = 5$. Before substituting these values let's also be sure we understand all the operations in the formula. Brackets, with no operation indicated, means the operation is multiplication. (This is just a convention that is agreed upon and not something you should be able to figure out.) For example, $(35)(21)$ actually means 35×21. There is also the operation of "exponentiation" or "raising to a power" in this formula. The expression 3^2 is called an exponential expression. We say three is raised to the second power. This is a shorthand for the product 3×3. That is, $3^2 = 3 \times 3 = 9$. In the same way,

$$3^4 = 3 \times 3 \times 3 \times 3 = 81$$
4 times

$$4^3 = 4 \times 4 \times 4 = 64$$
3 times

$$5^2 = 5 \times 5 = 25$$
2 times

Now we proceed to substitute into the formula and compute the total.

$$A = P\left(1 + \frac{r}{n}\right)^m \qquad \text{where } m = t \times n$$

$$= \$1000\left(1 + \frac{.06}{1}\right)^m \quad \text{where } m = 5 \times 1 = 5$$

$$= \$1000\,(1.06)^5$$

$$= \$1000 \times 1.06 \times 1.06 \times 1.06 \times 1.06 \times 1.06$$

$$A = \$1338.2255$$

So to the nearest penny, the total amount we obtain at the end of five years is \$1338.23. Therefore the interest will be $\$1338.23 - \$1000 = \$338.23$. As you can see when we compare this with the \$300 that would be earned in the simple interest account, the compound interest is a much better option.

Question 6
Explain each of the entries on a standard charge account billing form. The following sample will serve as an illustration for discussion.

Date	Store	Reference	Description	Charges	Payments or credits
6/26	BE	00124638	Mens Pants		16.74
6/26	BE	00042204	Cosmetics	13.58	
6/26	BE	00124641	Womens Pants		
7/09		00007018	Payment Thank you		164.04
7/24	SO	00885535	Charge	112.67	
7/24	SO	01001458	Mens Sweaters	36.64	

Previous Balance	Deduct Payments and Credits	Average Daily Balance for Finance Charge Only	Finance Charge 50¢ Minimum	Add Purchases	New Balance	Minimum Payment Due
255.46	180.78	151.79	2.28	177.11	254.07	▼

Closing Date This Month	Closing Date Next Month	Periodic Rate	Annual Percentage Rate			
7/24/81	8/24/81	1.5	18			30.00

It is surprising how many regular charge account customers never check the accuracy of their monthly billings. Yet errors in these bills, sometimes substantial ones, are known to occur. Most large accounts are now handled by computer. And although computerization of information is essential in our society today, the mechanics are by no means perfect. Machines still malfunction. Information is still fed into the machines by people and people make mistakes. We have all heard stories like the one about the customer who received a computerized bill for $250,000 to cover one month's electricity. Large errors like this are obvious immediately. We need to worry about smaller errors, ones that are easy to overlook unless the bill is checked carefully. In most established systems errors occur mainly in the entry of items into the system, rather than in performing the calculations. Computers are very reliable in doing the arithmetic. However, we will go over the important entries in the sample statement above, including the method of checking computations. Using your calculator to verify the arithmetic is good practice. Also, looking carefully at each entry, can increase our awareness of how to locate nonarithmetic errors.

Charges: First verify that all of the listed items were actually purchased and the amounts are correct. It is important to save sales receipts, particularly to check purchases like the one labeled only by "Charge" on 7/24.

Payments or Credits: All merchandise you have returned and the amount of any monthly payment you made should appear in this column.

Previous Balance: This is the balance in your account at the beginning of the bill period. It should be exactly the same figure as the NEW BALANCE from last month's bill.

Deduct Payments and Credits: This figure is unpleasant to verify yourself. Basically, it is found by adding up what you owed on each day and dividing by the number of days. You can at least check to see that it is within reason. It certainly should not be more than the sum of what you owed before (PREVIOUS BALANCE) and what you bought this month (ADD PURCHASES). If you have made a payment on the previous bill, that should substantially reduce the AVERAGE DAILY BALANCE. In this example, the sum of the previous balance and new purchases is: 255.46 + 177.11 = 432.57. There as a total reduction by payments and credits of 180.78. The difference of these two values is 432.7 - 180.78 = 251.79. Notice that the AVERAGE DAILY BALANCE (151.79) is even less than 251.79. Can you explain this? Observe that many of the new purchases were made late in the billing period, so these charges were not owed for the entire time. This is a useful piece of information, since you can minimize your AVERAGE DAILY BALANCE by timing your purchases within the billing period. As we will see in the next computation, this lowers the amount you pay in interest charges.

Finance Charge: Obtain this figure directly from the AVERAGE DAILY BALANCE by multiplying by the PERIODIC RATE. The PERIODIC RATE is given as 1.5 meaning that the monthly rate of interest is 1.5%. Notice that the annual percentage rate is given as 18, which is 1.5 X 12 (months). To calculate the finance charge we must convert the 1.5% to a decimal number by multiplying by .01.

$$
\begin{aligned}
\text{Finance charge} &= \text{Average daily balance} \times \text{Periodic rate} \\
&= 151.79 \qquad\qquad\qquad \times (1.5 \times .01) \\
&= 2.27685 \\
\text{Finance charge} &= 2.28 \text{ (to the nearest penny)}
\end{aligned}
$$

Add Purchases: This figure should be the sum of the entries in the "Charges" column.

New Balance: To find the new balance, add up the previous balance, purchases, and finance charge, and subtract your payments and credits. Using abbreviations we represent that formula below:

$$
\begin{aligned}
\text{New balance} &= \text{P.B.} \quad + \text{P.} \quad + \text{F.C.} - \text{P\&C} \\
&= 255.46 + 177.11 + 2.28 - 180.78 \\
\text{New balance} &= 254.07
\end{aligned}
$$

Minimum Payment Due: This figure depends on your NEW BALANCE and the type of charge account you have established with the company. It is either the total NEW BALANCE or some percentage of that figure.

CONCLUSION

That finishes our consideration of some sample situations in which you (and your calculator) can participate actively in mathematical computation. None of the examples were particularly earth shaking. The authors are confident you could survive adequately without concerning yourself with the details discussed. However, we do feel that an attitude of confidence in your ability to understand the mathematical details of everyday encounters is an important protection. It means you will feel free to question computations you don't understand, without feeling apologetic or embarrassed. It also means you can *choose* to do necessary calculations yourself if you feel the matter is important.

The following exercises provide some additional practice in calculator use. The first shows the actual cost of purchasing a house. The second provides practice in computing per-unit costs of grocery items. Solutions are shown at the end of the exercise.

Exercise: CALCULATOR USE

1. A house is purchased for the advertised price of $55,000. The purchase is made with a down payment of $14,000 plus a thirty-year mortgage. The monthly payments of $385 are applied to both repayment of the mortgage loan ($290) and to insurance and taxes ($95).
   a) Find the total annual cost of purchasing the home.
   b) Find the total amount paid annually on the loan itself.
   c) Find the total annual cost of insurance and taxes.
   d) Find the total amount paid on the loan after 30 years.
   e) Calculate how much is actually paid to buy the house (down payment and loan repayment).
   f) Find the total amount of interest paid in the purchase of the house.

2. On your next trip to the grocery store, take along your calculator and compute the following.
   From the price for a dozen eggs find:
   a) The cost of one egg; five eggs; three dozen eggs.
   b) Select a packaged item in the meat department. Somewhere on the label will be price per pound, weight of this piece, and total price. Verify that the total price is correct, using the other two figures.
   c) How much would the item in (b) cost if it was on sale for 10¢ less per pound?

    d) Most states suggest or require unit pricing on grocery items. Select a large can of tomatoes and from the information on the can compute the price per ounce. (If your store has unit pricing labels on the shelf you can verify your result there).

    e) Now as you actually shop, keep a tally of item costs. Compare it with the check out clerk's total (before any sales tax is added).

Solutions:

1a) (Total monthly payment) × (12 months) =
    $385   × 12   = $4620

b) (Taxes/insurance portion) × (12 months) =
    $95   × 12   = $1140

c) (Loan repayment portion) × (12 months) =
    $290   × 12   = $3480

d) (Annual loan repayment) × (30 years) =
    $3480   × 30   = $104,400

e) (Down payment) + (Loan payment) =
    $14,000 + $104,400   = $118,400

f) Interest = (Actual cost) − (Advertised price)
           = $118,400 − $55,000   = $63,400

(Notice the interest costs are more than the original price of the house!)

2a) Cost of one egg = (Price per dozen) ÷ 12.
    Cost of five eggs = (Price per dozen) ÷ 12 × 5.
    Cost of three dozen eggs = (Price per dozen) × 3.

b) Total price = (Price per pound) × (Weight of item)

c) Sale price = (Price per pound − .10) × (Weight of item)

d) Cost per ounce = (Cost of item) ÷ (Number of ounces)
Recall that 1 pound = 16 ounces; so if the can should say 1 pound, 13 ounces, the total number of ounces is 16 + 13 = 29.

FURTHER READING

Berman, Elizabeth. *Mathematics Revealed*. New York: Academic Press, 1979. (This is a summary of basic arithmetic)

Hughes-Hallett, Deborah. *The Math Workshop, Algebra*. New York: Norton, 1980.

Miller, Charles, and Vern Heeren. Consumer Mathematics. In *Mathematics an Everyday Experience*. Palo Alto, Calif.: Scott, Foresman and Company, 1976.

solutions to chapter 7 examples

APPENDIX A

1. 40 dogs in the kennel; answer was given in the first sentence of the problem.

2. Yes. The 4th of July is on the calendar in every country.

3. The surgeon is the boy's mother.

4. Change the orientation of the lake in order to double the area. Some elementary geometry is necessary to actually prove the new area is twice as large.

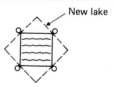

New lake

5. Abbott is the entomologist, Brown is the farmer, and Casper is the detective.

6. a) 16, 19   b) 32, 64   c) 13, 21   d) 36, 49

7. To find the next line add the two numbers adjacent and directly above.

$$\begin{array}{ccccccc} 1 & 6 & 15 & 20 & 15 & 6 & 1 \\ 1 & 7 & 21 & 35 & 35 & 21 & 7 & 1 \end{array}$$

8. a) Envelope: Milk goes into a glass, and a letter goes into an envelope.

   b) Triangle was compared to a square and if one is dotted then the other is solid.

   c) In all the other figures, the inner figure was a different shape than the outer.

   d) All the other figures have an outside shape that is made up of straight lines.

   e) The second figure is obtained from the first by holding the arrow fixed and rotating the rest of the figure about the point where it attaches.

   f) Flip figure over and then open up the inner part.

9. 216 miles. Another, faster method uses a proportion. We must be sure to convert all times to the same units. In this solution the times are expressed in minutes. 4 hours is $(4 \times 60) + 30 = 270$ minutes. M stands for number of miles traveled.

$$\frac{M}{270} = \frac{36}{45}$$ Now multiply both sides by 270.

$$M = \frac{36 \times 270}{45}$$ Perform the operations on the right.

$$M = 216$$

10. 1024

11. See sample solution given in text. Many other solutions would also be correct.

12. Think of the balls as being numbered 1 through 8. All alternatives on each weighing must be considered.
   First Weighing: Put 1, 2, and 3 on one side of scale and 4, 5, and 6 on the other. Either they balance or one side goes down.
   Case I: They balance; the odd ball is then either 7 or 8.
   Second Weighing: Put 7 on one side and 8 on the other. One side must go down. It contains the odd ball.
   Case II: They don't balance; we can conclude the odd ball is on the heavy side (assume it is the side containing 1, 2, and 3).
   Second Weighing: Put 1 on one side and 2 on the other. Either they will balance on this weighing or one side will go down.

If they balance: Then 1 and 2 are okay so 3 is the odd ball.
If they don't balance: Then the side that is heavier contains the
odd ball.

13.   The rectangle that measures 5 X 5, giving an
area of 25 square feet is the largest. Notice the
perimeter adds up to the 20 feet of fencing.

14. The cars will be at the same spot at the end of 2 hours. In that
time Car B will have gone 120 miles. Car A will only have gone
100 miles but it began 20 miles ahead.

15. 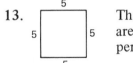   The key to the solution is working in three dimensions. Four triangles can be made with 6 sticks, if
they are arranged in the form of a pyramid with
triangular base, called a tetrahedron.

16. Many reasonable estimates could be given. Don't be reluctant
to make some guess, even if you are totally unfamiliar with the
mathematical formulas involved. You could certainly be sure
that $0.00 would be a low estimate and say $1,000 is high.
Using the formula that will be developed in Chapter 10 we can
show that the actual value is $112.65.

17. The high estimate should be no more than $39.95, the original
price. A low estimate might be half of that or about $20. The
actual cost is .85 X $39.95 = $33.96.

18. The article costs 17¢. Begin by considering all the possible
combinations of two coins necessary to purchase three of the
items. The standard coins are: 1¢, 5¢, 10¢, 25¢, 50¢, and 100¢.
The following is a list of all the totals that could be obtained
by adding two of these coins together.

1¢ + 1¢ = 2¢	5¢ + 5¢ = 10¢	10¢ + 10¢ = 20¢
1¢ + 5¢ = 6¢	5¢ + 10¢ = 15¢	10¢ + 25¢ = 35¢
1¢ + 10¢ = 11¢	5¢ + 25¢ = 30¢	10¢ + 50¢ = 60¢
1¢ + 25¢ = 26¢	5¢ + 50¢ = 55¢	10¢ + 100¢ = 110¢
1¢ + 50¢ = 51¢	5¢ + 100¢ = 105¢	
1¢ + 100¢ = 101¢		25¢ + 25¢ = 50¢
		25¢ + 50¢ = 75¢
	50¢ + 50¢ = 100¢	
100¢ + 100¢ = 200¢	50¢ + 100¢ = 150¢	25¢ + 100¢ = 125¢

We can eliminate any total that is not evenly divisible by three since it should be the cost of three items. The totals that are underlined are those that are divisible by three. This reduces the possibilities to only eight combinations of coins. Each of these must now be checked against the first statement of the problem, namely that one of the items could be purchased with a *minimum* (meaning no less than) of four coins. Six of the eight possibilities are shown below:

Cost of three	Cost of one	Minimum number of coins
6¢	2¢	2 coins (1¢ + 1¢)
51¢	17¢	4 coins (10¢ + 5¢ + 1¢ + 1¢)
15¢	5¢	1 coin (5¢)
30¢	10¢	1 coin (10¢)
105¢	35¢	2 coins (25¢ + 10¢)
60¢	20¢	2 coins (10¢ + 10¢)

Two other totals have not been included in the table. Can you fill in the lines for 75¢ and 150¢? It will be seen that only one of the eight alternatives satisfies the first statement. That is when a single item costs 17¢. Finally, the second condition must also be verified. Two items would cost 17¢ × 2 = 34¢. To make up 34¢ requires combining a *minimum* of 6 coins, namely 25¢ + 5¢ + 1¢ + 1¢ + 1¢ + 1¢.

solutions to
chapter 8 problems

APPENDIX B

READING PROBLEMS

1. Nine cows are left.

2. The answer is 12. All of the months have 28 days.

3. The first thing to light would have to be the match.

4. Roosters don't lay eggs.

5. If the man has a widow, then he is not alive to marry anyone.

6. The survivors probably wouldn't appreciate being buried in either country.

7. There is no missing dollar. The entire original amount ($30) is acounted for this way:

Clerk has	$25	
Bellhop has	2	the $27 paid by the three women
Women have	3	
Total	$30	

The erroneous explanation is confused because it adds the $2 the bellhop has, to the $27 the women paid. The $2 is actually part of the $27 paid. See above.

8. There are 12 stamps in a dozen (of any denomination).

SPATIAL PROBLEMS

9.

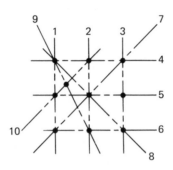

10. There are many different ways to cut up the cube. The fewest number of cuts that would be necessary is six. See sketch in Problem 11.

11. 3 red sides—8 cubes—from the corners.
    1 red side—6 cubes—from the center of each face.
    2 red sides—12 cubes—from the center of each edge.
    4 red sides—none
    0 red sides—1 cube—from the very center of the original cube.

12.

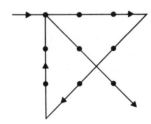

13. Rotate one-half of the way around (180°).

14. Rotate one-quarter of the way around (90°) and then flip over.

15. Their right feet will never strike the ground at the same time.

16. **a)** perpendicular   **b)** parallel

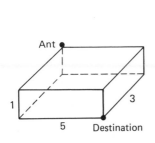

17. The ant travels five feet. The solution to this problem depends on changing the frame of reference from three dimensions to two dimensions. This is done by flattening out the box, or at least visualizing the box as flattened.

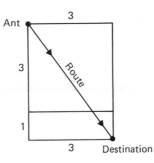 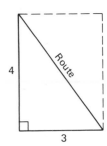

Now the length of the route is found using the Pythagoran theorem as follows:

$$
\begin{aligned}
\text{Route} &= \sqrt{4^2 + 3^2} \\
&= \sqrt{16 + 9} \\
&= \sqrt{25} \\
&= 5
\end{aligned}
$$

18. There would be 3 inches less than 2 yards.

19. The bear would be white since the hunter would have to be at the North Pole.

20.

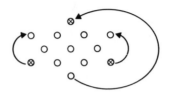

21. a) 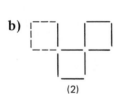 Remove the sticks shown as dotted lines leaving one large square with a smaller one inside.

(1)

b) Sticks shown as dotted lines in square (1) are removed and placed as shown in square (2).

(2)

22. The straight cuts should be made on the indicated dotted lines and piece relocated as shown.

a)   (b)

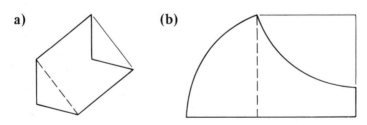

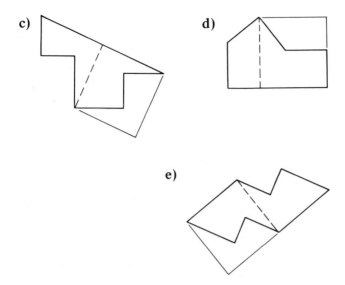

23. 2 minutes; 1 minute to get the engine through the tunnel and another minute to get the caboose through.

24. 40 miles. The riders meet in exactly one hour at the speeds they are traveling. In that time the fly travels 40 miles regardless of which direction it is going during any part of the time.

25. Put on the scale: 1 bar from stack one, 2 bars from stack two, 3 bars from stack three and so on. There will finally be 10 bars from stack ten. Find the total weight to expect on the scale if all of the bars weighed a full pound. $1 + 2 + 3 + 4 + 5 + 6 + 7 + 8 + 9 + 10 = 55$ pounds. Now read the actual weight on the scale. If the result is 1 ounce under 55 lbs., then the odd stack is the first one. If the scale reads 2 ounces under 55 lbs., the odd stack is the second, and so on.

26. The beggar is the woman's sister.

27. Four socks

28. She has on a black hat. Below are sketched the four possibilities for the colors of the hats on the first two women in line. "B" stands for black hat and "W" stands for white hat. The comments made by the last and middle women in line eliminate all but the third and fourth alternatives. Both of these have a black hat on the first woman.

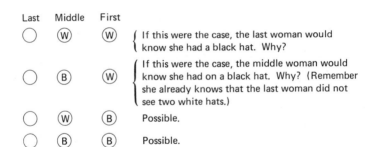

Last	Middle	First	
◯	Ⓦ	Ⓦ	{ If this were the case, the last woman would know she had a black hat. Why?
◯	Ⓑ	Ⓦ	{ If this were the case, the middle woman would know she had on a black hat. Why? (Remember she already knows that the last woman did not see two white hats.)
◯	Ⓦ	Ⓑ	Possible.
◯	Ⓑ	Ⓑ	Possible.

29. First native—Bad;
    Second native—Good;
    Third native—Good;
    Fourth native—Bad.

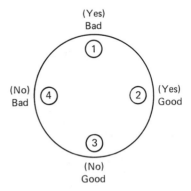

30. 3/4 of the job.

31. 1949

32. Will vary depending on the year you are working in.

33. **a)** Divide the remainder equally, since they have both eaten 1/3 of the pizza.
    b) Each should pay the same amount, $1.80.

34. **a)** Brad is the shortest.
    b) It is either Jim or Mary, but there is not enough information given to determine which is the case.
    c) Jim is tallest.

35. Jim has 6 × 2" = 12".
    Brad has 1/2 × 12" = 6".
    Sue has 1/2 × 6" = 3".

36. There are two ways to answer each question. Since there are five questions on the test there will be 2 × 2 × 2 × 2 × 2 = 32 ways to answer the test.

37. Joe has $5 and Kathy has $20.

38. The following summarizes the results.

Father's last name	Butcher	Baker	Tailor	Carpenter
Father's occupation	carpenter	butcher	baker	tailor
Son's occupation	baker	tailor	carpenter	butcher
Wife's maiden name	Tailor	Carpenter	Butcher	Baker

39. The following chart summarizes the results.

House color	yellow	blue	red	ivory	green
Nationality	Norwegian	Ukranian	English	Spaniard	Japanese
Beverage	CUTTY SARK	tea	milk	O.J.	coffee
Cigarette	Kools	Chesterfields	T.G.	Camels	L.S.
Animal	fox	horse	snails	dog	ZEBRA

40. Mildred—40—Top floor
    Elden—50—Third floor
    Rita—45—Second floor
    Wes—25—First floor

bibliography

APPENDIX C

Adams, James L. *Conceptual Blockbusting, A Guide to Better Ideas*. New York: W.W. Norton Pub., 1974.

Anderson, Harold H. (ed.) *Creativity and Its Cultivation*. New York: Harper and Brothers, Publishers, 1959.

Beck, Aaron T. *Cognitive Therapy and the Emotional Disorders*. New York: The New American Library, 1976.

Bell, Eric. *Men of Mathematics*. New York: Simon and Schuster, 1937.

Benson, Herbert. *The Relaxation Response*. New York: Avon Books, 1975.

Berman, Elizabeth. *Mathematics Revealed*. New York: Academic Press, 1979.

Bloom, Lynn Z., Karne Coburn and Joan Pearlman. *The New Assertive Woman*. New York: Dell Publishing Co., 1970.

Campbell, David. *Take the Road to Creativity and Get Off Your Dead End*. Illinois: Argus Communications, 1977.

Chamberlain, Jonathan M. *Eliminating a Self-Defeating Behavior*. Provo, Utah: Brigham Young University, 1974.

Cudney, Milton R. *Eliminating Self-Defeating Behaviors*. Kalamazoo, Mich: Life Giving Enterprises, Inc., 1975.

Donady, Bonnie, Stanley Kogelman and Sheila Tobias. "Math Anxiety and Female Mental Health: Some Unexpected Links," (unpublished), Wesleyan University, Middletown, Conn., August 4, 1976.

Fensterheim, Herbert, and Jean Baer. *Don't Say Yes When You Want to Say No.* New York: David McKay Co., Inc., 1975.

Frand, Jason L. *How to Study Mathematics, Chemistry, Statistics, Physics.* Culver City, Calif.: SKIL Publishing Co., 1979.

Galarosa, Annie, Ann Oxrieder and Marilyn Weckwerth. "How to Take Tests." Seattle: Seattle Central Community College, 1975.

Hill, Joseph E. "The Educational Sciences." Bloomfield Hills, Mich. Oakland Community College, 1971.

Holleran, Bruce P., and Paula R. Holleran. Creativity Revisited: A New Role for Group Dynamics. *The Journal of Creative Behavior* 10 (2): 130, 1976.

Holmes, T.H., and R.H. Rahe. The Social Readjustment Rating Scale. *The Journal of Psychosomatic Research.* 11:213–218, 1967.

Hughes-Hallett, Deborah. *The Math Workshop, Algebra.* New York: W.W. Norton Publishers, 1980.

Koestler, A. *The Act of Creation.* New York: Macmillan Co., 1964.

Kogelman, Stanley, and Joseph Warren. *Mind Over Math.* New York: McGraw-Hill Book Company, 1978.

Lakein, Alan. *How to Get Control of Your Time and Life.* New York: New American Library, 1973.

LeShan, Lawrence. *How To Meditate.* New York: Bantam Books, 1974.

Miller, Charles, and Vern Heeren. *Mathematics an Everyday Experience.* Palo Alto, Calif.: Scott, Foresman and Co., 1976.

Mooney, Ross L., and Taher A. Razik (eds.) *Explorations in Creativity.* New York: Harper and Row, Publishers, 1967.

Osborn, Alex. *Applied Imagination.* New York: Charles Scribner's Sons, 1957.

Oxreider, Ann, Mike Linscott, Sanford Helt and Rafael Marino. "How to Read a Math Textbook." Seattle: Seattle Central Community College, 1978.

Parnes, Sidney J., Ruth B. Noller and Angelo M. Biondi. *Guide to Creative Action.* New York: Charles Scribner's Sons, 1977.

Remick, Helen. Participation Rates in High School Mathematics and Science Courses. *The Physics Teacher* May 1978.

Resnikoff, H.L., and R.O. Wells, Jr. *Mathematics in Civilization.* New York: Holt, Rinehart and Winston, 1973.

Scheerer, Martin. Problem-Solving. *Scientific American*. 208 (4): 118-128, 1963.

Sells, Lucy. "High School Mathematics as the Critical Filter in the Job Market."
    In Developing Opportunities for Minorities in Graduate Education, Proceedings
    of the Conference on Minority Graduate Education at the University of
    California, Berkeley, May 1973, pp. 47-59.

Shaw, Harry. *Thirty Ways to Improve Your Grades*. New York: McGraw-Hill Book,
    Co., 1976.

Smith, Manuel, J. *When I Say No I Feel Guilty*. New York: Dial Press, 1975.

"The Key Word Approach to Note Taking," available through Seattle Central
    Community College, Seattle, Wash.

Tobias, Sheila. *Overcoming Math Anxiety*. New York: W.W. Norton and Company,
    Inc. 1978.

Weckwerth, Marilyn. *Stress Management*. Seattle: North Seattle Community
    College, 1979.

Whimby, Arthur, and Jack Lochhead. *Problem Solving and Comprehension*.
    Philadelphia: The Franklin Institute Press, 1980.